D1204425

Evolutionary Trends in Flowering Plants

Evolutionary Trends in Flowering Plants

Armen Takhtajan

 Columbia University Press NEW YORK

Columbia University Press
New York   Oxford

Copyright © 1991 Columbia University Press
All rights reserved

Library of Congress Cataloging-in-Publication Data

Takhtadzhian, A. L. (Armen Leonovich)
  Evolutionary trends in flowering plants / Armen Takhtajan.
    p.   cm.
  Includes bibliographical referencesand index.
  ISBN 0-231-07328-3 (acid-free paper) : $40.00
  1. Angiosperms—Evolution.   2. Plants—Evolution.   I. Title.
QK495.A1T35   1991
582.13'0438—dc20   91-7320
                                                           CIP

Casebound editions of Columbia University Press books are Smyth-
sewn and printed on permanent and durable acid-free paper

⊚

Printed in the United States of America

c 10 9 8 7 6 5 4 3 2 1

93-0748 NCL

Contents

Preface ix

Introduction 1
    1. Modes of Evolutionary Alterations of Ontogeny 2
        1.1. Prolongation (Additions) 3
        1.2. Abbreviation 4
        1.3. Deviation 6
        1.4. Combination of Terminal Abbreviation with
            Deviation (Neoteny) 7
    2. Significance of "Living Fossils" 10
    3. Evolutionary Series of Characters (Morphoclines) 12

1. Evolutionary Trends in Vegetative Organs 21
    1.1. Growth Habit 21
    1.2. Branching 26
    1.3. Evolutionary Trends in Leaves 28
        1.3.1. Evolutionary Trends in Leaf Form 29
        1.3.2. Evolutionary Trends in Leaf Venation 30
        1.3.3. Evolutionary Trends in the Structure of
            Minor Veins 42
        1.3.4. Leaf Vernation and Leaf Arrangement 45
    1.4. Stomatal Apparatus 45
    1.5. Nodal Structure 47
    1.6. Evolutionary Trends in Tracheary Elements of
        Axial Organs 51
        1.6.1. Origin and Evolution of Vessels 52
        1.6.2. Origin and Evolution of Sieve Tubes 58

1.6.2. Origin and Evolution of Sieve Tubes   58
1.6.3. Evolutionary Trends in Radial and Axial Parenchyma
of Secondary Xylem and Phloem   61
1.6.4. Origin and Evolution of Wood Fibers   65

2. *Evolutionary Trends in Flowers and Inflorescences*   75

2.1. General Floral Structure   75
2.2. From Spiral to Cyclic Flowers   76
2.3. Oligomerization of the Homologous Flower Parts   78
2.4. Origin and Evolution of the Perianth   79
2.5. Evolutionary Trends in Stamens and Androecium   82
2.5.1. Stamens   82
2.5.2. Androecium   86
2.6. Evolutionary Trends in Carpels and Gynoecium   90
2.6.1. Initial Stages of Evolution of Carpels   90
2.6.2. Evolution of the Gynoecium   96
2.6.3. Evolution of Placentation   104
2.6.4. Origin of the Inferior Ovary   109
2.7. The Evolution of Inflorescences   112

3. *Microsporangia, Microspores, and Pollen Grains*   135

3.1. The Microsporangium   135
3.2. Microsporogenesis and Microspores   138
3.3. Pollen Grains   139

4. *The Ovule, Megasporangium and Megaspores*   159

4.1. Origin of the Integument   160
4.2. Origin of the "Double" Integument   161
4.3. Form and Orientation of Ovules   164
4.4. Evolution of the Megasporangium: The Megaspore   166

5. *Evolutionary Trends in Pollination*   171

6. *Evolution of Male and Female Gametophytes: Fertilization and Triple Fusion*   185

6.1. Origin of the Male Gametophyte   186
6.2. Origin of the Female Gametophyte   188

6.3. Penetration of the Pollen Tube, Fertilization, and
Triple Fusion 194

7. *Evolution of Fruits 199*

7.1. Apocarpous Fruits 200
7.2. Dry Syncarpous Fruits 204
7.3. Fleshy Syncarpous Fruits 208

8. *Evolution of the Seed 213*

8.1. Emergence of the Monocotyledonous Embryo 214
8.2. Evolutionary Trends in Endosperm Formation 216
8.3. Types of Endosperm Specialization 218
8.4. The Seed Coat 219
8.5. Origin of the Fleshy Seed Appendages 222

9. *Mosaics of Evolutionary Trends and Heterobathmy of Characters 227*

Index 233

Preface

I have long been interested in various aspects of evolutionary morphology of flowering plants, especially in transformation series of morphological characters. In 1948, I outlined evolutionary morphology in a book entitled *Morphological Evolution of the Angiosperms,* which was published in Russian. In 1959, in my German book *Die Evolution der Angiospermen,* I returned to the subject with some necessary alterations and additions. Finally, in 1964, I published in Russian *Foundations of the Evolutionary Morphology of Angiosperms,* which was an abridged and updated version of the first book. More than 25 years have elapsed since the appearance of that book and so much new and important data have been published since then that I felt a need to write a new overview of the whole subject. In this new book, I have tried to concentrate chiefly on the main evolutionary trends, emphasizing those characters that are of a special systematic and evolutionary significance. I have not described many important morphological details because the reader can easily find them in current textbooks of plant morphology, anatomy, embryology, and palynology. This helped me to make the text as concise and condensed as possible. In order not to make this book overwhelmingly large, I have restricted discussions of a number of theories and hypotheses. I have not touched at all on trends in biochemical evolution, evolution of karyotype, or sieve-element plastids. The excuse is both the limitations and scarcity of the knowledge and the capacities of the author. Yet, it seems to me that the study of evolutionary trends in flowering plants is so essential for evolutionary botany as well as for phylogeny and systematics in particular, that even a relatively less comprehensive treatment of this important subject is justified.

The writing of the book took place at the Missouri Botanical Garden in St. Louis, and I thank Dr. Peter H. Raven, the Director, for providing a congenial atmosphere and the Gardens fine facilities. I owe especial thanks to Dr. Nancy Morin, Missouri Botanical Garden, for taking care of the matters connected with the publication of the book. It is also a pleasure to thank Dr. Arthur Cronquist, New York Botanical Garden, for reading the whole manuscript, and Dr. Peter Bernhardt, St. Louis University, for reading the chapter on pollination. Both of them made valuable suggestions that were very helpful during the preparation of the final version of the book. Many thanks also to Eloise Cannady, Missouri Botanical Garden, for the careful typing of the manuscript, and to Dr. Dale Johnson, Library, Missouri Botanical Garden, for his help with the bibliography. And last but not least, I thank my wife Alice for her continuing aid in many ways.

Evolutionary Trends in Flowering Plants

Introduction

Evolutionary morphology, like any other science, cannot restrict itself to the collection of facts alone, and it cannot collect them without selection, considering them as equally important, as well as without analysis and generalizations. There was—and there still is—a real danger that the facts will continue to accumulate faster than they can be analyzed and generalized. In the past, a multitude of factual material was published in purely descriptive works but much of this material was not analyzed in time, became outdated, and now only adds dead weight to the literature. Many purely descriptive works which have not been analyzed are almost of no interest to modern science. It should be recognized that there is nothing more ephemeral than purely descriptive research. But among empirically oriented researchers, the myth is prevalent that a true scientist proceeds from the observation of facts without any preliminary concepts and hypotheses. In the philosophical literature this myth has been called the "Fallacy of tabula rasa." As is well known, however, the description of facts without preliminary ideas and concepts is logically impossible. Without preliminary ideas, it is impossible to know just what facts should be described, what in these descriptions is of more importance for research, and what is of minor importance.

Placing facts above ideas, which is characteristic of extreme empiricism, has an injurious influence on the development of plant morphology. The facts by themselves, however accurately they may be described, are only the raw material of science. They should be interpreted, systematized, and generalized. The spirit of generalization reigns in science. Thus, the main goal of science cannot be the chance collection of facts unconnected with specific problems and aims. The collection of facts always requires some hypothesis or the other. Without a theory, we do not even know what in particular is to be observed and described. The observations and collection of information should be programmed on the basis of some ideas and hypotheses, however preliminary they may be. As Koltzoff put it: "it is better to work with a bad hypothesis which can be disproved than without any hypothesis, when it is not known what is to be proved or disproved (1936:648)." Or as Darwin wrote many years ago in one of his letters "all observation must be for or against some view (F. Darwin and Seward 1903:195)." The relation between observation and theory has been deeply analyzed by Karl Popper (see especially Popper 1965 and 1975).

1. Modes of Evolutionary Alterations of Ontogeny

In evolution, hereditary changes in structures manifest themselves at the most diverse stages of their morphogenesis, starting with the formation of primordia and ending with the last developmental phases. This idea, expressed independently by various authors in the last century, has gained its recognition only in the twentieth century. But it has developed almost exclusively in the morphology of animals, where it is widely employed. In botanical morphology, however, the "theory of phylembryogenesis," as it was called by Sewertzoff (1927, 1939), was not adopted for a long time, though some similar ideas were repeatedly expressed. In 1943, I attempted to use this idea in plant morphology and later repeatedly returned to this subject (see especially Takhtajan 1972). However, the theory of "phylembryogenesis" penetrates botany with difficulty. Like other general principles, it frequently runs against a kind of mental idiosyncrasy.

Any evolutionary change in adult structure of organisms is the result of hereditary alterations in ontogeny in successive generations. Evolutionary alteration of organisms may occur at the most diverse stages of ontogenetic development. All these alterations constitute greater or lesser deviation of ontogeny from its previous course. As a result of ontogenetic alterations there arise more or less substantial changes in the adult structure of either distinct parts of organs, or the entire organism. The nature and scope of these alterations depend on the mode of "phylembryogenesis." This theory has been worked out by Sewertzoff (1927, 1931), Franz (1927), Remane (1956), De Beer (1958), Rensch (1959) and others. Taking these works into account we can classify possible patterns or modes of ontogenetic changes in evolution of higher plants as follows:

1.1. Prolongations (Additions)

1. Terminal prolongation (Sewertzoff's "anaboly" or "superposition of stages")—addition of new stages to the final phases of development.

Terminal prolongation takes place easily and occurs in numerous cases. Through terminal prolongation various structures of pollen grains, seed-coat, pericarp, and various parts of flower appear, especially all types of outgrowths. In fruiting, the sepals of the Dipterocarpaceae expand into wings, a phenomenon which is a typical terminal prolongation. Terminal prolongations occur widely in the plant world and may be observed frequently. Prolongation is usually more or less gradual in its character and does not produce any new evolutionary novelties of great importance. In spite of the numerous adaptations which are brought about by terminal prolongation, its creative potentialities are limited. Another characteristic feature of terminal prolongation is the possibility of "recapitulation" or repetition of ancestral adult stages (see Sewertzoff 1927, Gould 1977).

2. Medial prolongation—intercalation of new stages in the middle of development.

Medial prolongation is much less common in the plant world than terminal prolongation. The insertion of a new intermediate stage meets many more difficulties from the morphogenetic point of view than terminal prolongation. Nevertheless, in a number of cases, intercalation has been of great importance for the evolution of higher plants. The intercalary meristem which appears as inserted zone of growth may serve as a typical example of medial prolongation. The intercalary meristem is characteristic of the sporophyte of the Anthocerotales and of the internodes of Equisetales, as well as as the stems of some liliopsids (mainly grasses) and the leaves of flowering plants.

3. Basal prolongation—addition of new stages at the very early phases of development.

Basal prolongation is mentioned here only as a theoretical possibility. I find it difficult to give any evident example of basal prolongation.

1.2. Abbreviation

1. Terminal abbreviation—omission of the final stages of development.

Abbreviation is directly opposed to prolongation. The most common case is the omission of the final stage of development. All types of vestigiation usually begin by way of terminal abbreviation. Thus, reduction of the corolla, androecium, and gynoecium may take place by way of progressive terminal abbreviation, when the final stages of development drop out one after another until the process of vestigiation is over with complete "aphanisy," i.e., disappearance of the organ. In a similar way, vestigiation of leaves, roots, and other organs may take place. However, by means of terminal abbreviation occur not only vestigiation and disappearance of organs but also their evolution in a new direction: the terminal abbreviation lies at the base of neoteny (see further).

2. Medial abbreviation—omission of the intermediate stages of development.

Though medial abbreviation is met with quite often, it is much less significant. Its role lies in shortening (and consequently, in accelerating) development through the exclusion of those intermediate stages that have lost their importance and become unnecessary. Thus, the scalariform tracheids typical of the metaxylem of the cordaits do not develop at all in the wood of the conifers. The stage of scalariform tracheids has dropped out from the ontogeny of the conifers. Medial abbreviation led to the bisporic and tetrasporic types of angiospermous female gametophyte. It played a certain role also in origin of the male gametophytes of Cycadales and *Gnetum* (excalation of the second prothallial cell). Medial abbreviations are very characteristic of angiospermous flowers.

3. Basal abbreviation—omission of the earlier stages of development.

Basal abbreviation occurs comparatively rarely. But, in some cases, it may play a considerable role in evolution. By way of a typical basal abbreviation, i.e. by a dropping out of a stage of prothallial cells, the male gametophyte of the *Taxus* type appeared. The same happened in the flowering plants, but in addition they underwent the terminal abbreviation (see chapter 6).

1.3. *Deviation*

1. Terminal deviation—deviation of the last stages of development from its previous course.
2. Medial deviation (Remane's "mesoboly")—deviation of the intermediate stages of development.
3. Basal deviation (Remane's "archiboly")—deviation of the earlier stages of development.
4. Total deviation (Sewertzoff's "archallaxis")—general deviation of the whole ontogeny as a result of abrupt changes of initial stages.

Deviation includes all sorts of divergence from the previous course of development of the entire organism or of its parts. It may arise at any stage of development. Like other evolutionary changes in ontogeny, deviation occurs more easily and frequently in later stages of development. Evolutionary alterations of the final developmental stages lead to the least significant deviations from the previous course of development. Therefore, the less significant the deviation is, the later are the stages during which it occurs. Relatively smaller deviations are usually realized in the terminal stages of development. For example, the relative dimensions of parts and their mutual arrangement including many changes in symmetry may be affected at the later stages of morphogenesis. Thus, formerly symmetrical leaves may acquire an asymmetrical form by the end of morphogenesis, and an actinomorphic flower may turn into a zygomorphic type. Splitting of the flabellate and pinnate leaves of a good many palms also occurs at the last stages of morphogenesis.

Having started from the last stages of morphogenesis, deviation may gradually affect even the earliest stages of development. It is exactly in this manner that many major evolutionary transformations of organs usually take place. By way of medial and—particularly—basal deviation, radical alterations in the structure of leaves, sporophylls, strobiles, flowers, etc., take place.

The more significant an evolutionary alteration of ontogeny is, the earlier are the stages during which it is brought about. Total deviation of ontogeny from its previous course occurs only through an abrupt and sharp "macromutational" alteration of the initial stages. In the structural evolution of plants and animals, there are many such changes that could appear in no other way except as sudden and discontinuous macromutational alterations in the course of development or Sewertzoffian "archallaxis." Therefore, the number of all kinds of symmetrically arranged structures, for example the number of leaves in a whorl or the number of sepals, petals, stamens, or carpels in the flower with a cyclic arrangement of the parts might have changed only through archallaxis. Thus, the tetramerous perianth usually originated from the pentamerous (or from the trimerous in some cases) neither by reduction of one of the members of each whorl nor by the concrescence of two of the five members of each circle but by a sharp change in the number of primordia, i.e. through archallaxis. Later, we shall again come across a number of glaring examples of the role of total deviation in the morphological evolution of flowering plants.

1.4. Combination of Terminal Abbreviation with Deviation (Neoteny)

In the evolution of the flowering plants as well as in that of the entire organic world (including the origin of mankind), the combination of terminal abbreviation and deviation is of prime importance. This extremely significant mode of morphological evolution is well known under various names, of which neoteny is the most commonly used. With many other authors (including Wardlaw 1952; Davis and Heywood 1963; Stebbins 1974;

and Corner 1976), I use the evolutionary term "neoteny" in its broader meaning for any truncation of ontogeny and premature completion of development of the whole organism (sporophyte or gametophyte) or any parts of it, that is, for a genetically controlled extension of the earlier phases of development into maturity, the former adult phase being omitted from the ontogeny. As a result previous stages of development are turned into the adult stages of the neotenical derivatives. This "Peter Pan" evolution includes both Kollmann's neoteny sensu stricto and Giard's and Gould's progenesis (see Gould 1977) as two different modes of hereditary juvenilization. The term "neoteny" is not a very felicitous one and there are a number of more or less complete synonyms. The terms "paedomorphosis" (Garstang 1922) and especially "juvenilization" (Huxley 1942) perhaps convey most exactly the content of the concept, but the term "neoteny" is so widely used in the literature that there is hardly any sense in replacing it.

The significance of neoteny for rapid and profound evolutionary changes depends on the simplification and despecialization of the neotenic organisms or their parts. Neotenous "rejuvenilization" increases evolutionary plasticity and opens new evolutionary avenues. "It is this possibllity of escaping from the blind alleys of specialization into a new period of plasticity and adaptive radiation which makes the idea of paedomorphosis so attractive in evolutionary theory," says Huxley (1954:20). Hardy (1954:128) comes to an analogous conclusion. The genetic basis of this increase of evolutionary plasticity of "juvenilized" organisms or their parts lies in the fact, long ago indicated by Koltzoff (1936:520), that abrupt neoteny involves at first great simplification of the phenotype alone, whereas the genotype maintains its complexity. Conservation of the former rich reservoir of genes that are not manifest in the development of the neotenous forms (but which are able to mutate into new active genes) leads to a high degree of their variability "and sometimes enable them to display an exuberant outburst of further progressive evolution" (Koltzoff 1936:520). Even single mutations with phenotypic effects large enough to alter the course of development would drastically change ontogeny and initiate neotenous transformation. Therefore, neoteny is basically a macroevolutionary process (Takhtajan 1983).

While the important evolutionary role of neoteny has been appreciated by many zoologists (see especially Garstang 1922, Koltzoff 1936, Remane 1956, Hardy 1954, De Beer 1958), only very few botanists concede a certain role of neoteny in the origin and evolution of higher taxonomic groups. Botanists usually attach only a secondary role to neoteny in the origin of certain species, more rarely of genera (e.g., *Phylloglossum* and *Welwitschia*), and very rarely of families (e.g., Lemnaceae) (for details see Vassilczenko 1965). At any rate, no botanist has applied this concept on the same large scale as has been done by zoologists, though some botanists like Agnes Arber (1937, 1950) attach some importance to neoteny in plant evolution.

In a series of publications starting in 1943 and summarized in 1976, I attempted to develop the concept of neoteny on the botanical material. I explained some macroevolutionary events in the history of the plant world on the basis of this concept. Thus, I put forward the opinion that the appearance of some large and successful groups of plants, including Magnoliophyta, is the result of neotenic mode of evolution. I applied the concept of neoteny in explaining the origin of herbaceous magnoliopsids from woody ancestors, the origin of liliopsids, as well as the origin of the flower, the male and female gametophytes, and some other organs and structures (see the following chapters).

2. *Significance of "Living Fossils"*

Evolutionary morphology and phylogeny of many extant groups of vascular plants are based on correlated studies of both fossil and living forms. However, flowering plants occupy, in this respect, a different position. The initial stages of angiosperm evolution are completely unknown to us: there is not yet any fossil record of the earliest magnoliophytes and their immediate ancestors. Besides, the fossil flowering plants are represented almost exclusively by separate remains of various vegetative organs (mainly leaves) and usually dispersed pollen grains, less frequently by fruits and seeds, and only rarely by flowers and their parts. The preservation of flowers in fossils is a very rare palaeobotanical event and it is therefore not surprising that we know nothing about pre-Aptian flowers and there are found only a few flowers in the rocks of Aptian age (Taylor and Hickey, 1990). Thus, very little is known about the Early Cretaceous flowers and nothing about the flowers of the Barremian age. In spite of great progress of fossil botany of flowering plants during the last decades, it provides only very scanty data from which conclusions on their structural evolution may be drawn. I concur with Stevens that "Fossil evidence is too sketchy to have affected ideas on angiosperm phylogeny deeply (1980:342)." However, for the study of evolutionary morphology and phylogeny, flowering plants have some advantages over the other vascular plants due to their being comparatively younger. While the initial forms of gymnospermous plants (seed ferns) became extinct long ago, many undoubtedly archaic flowering plants with a number of primitive characters such as Degeneriaceae, Magnoliaceae, Winteraceae, and some others, are still preserved as "living fossils." The study of "living fossils" is of fundamental importance—to a considerable extent it compensates the insufficiency of palaeobotanical record. It gives a chance to trace the evolution of certain organs and tissues starting from the early stages of their origin and, in some cases, even to observe them in statu nascendi. In archaic groups, morphological structures and functions are usually less obscured by the processes of specialization and reduction and therefore yield more readily to evolutionary interpretation. For instance, the Winteraceae and Degeneriaceae provide us with much more information about the morphological nature of carpel and the origin of stigma than any other advanced groups. This is equally true for all other morphological structures, both vegetative and reproductive. Many intricate problems of angiosperm morphology become more lucid in the light of our knowledge of the morphology of the archaic groups.

However, we should not interpret primitiveness of structural characters of archaic groups too straightforwardly. We should always remember that all of them are only ancient side branches of the evolutionary tree and there are no truly ancestral forms among them. The evolutionary process like any other historical processes is not parsimonious (Cain 1982, Friday 1982); it did not follow "Ockham's Razor." One can even say that the evolutionary process, especially in earlier stages of cladogenesis, is characterized by a considerable degree of redundancy. As a result of exuberant cladogenesis at the early stages of angiosperm evolution, there emerged many "experimental models" most of which became extinct. What we have now are no more than insignificant remnants of a great diversity of evolutionary endeavours. One should therefore be very cautious when interpreting morphological characters of these "living fossils." Some of their characters are undoubtedly very primitive, but some others are marks of ancient specializations.

3. *Evolutionary Series of Characters (Morphoclines)*

One of the main tasks of evolutionary morphology is to ascertain evolutionary sequences of characters, their continuous transformation series or morphoclines (Engler 1892; Hallier 1912; Bessey 1915; Sprague 1925; Takhtajan 1947; Maslin 1952; Sporne 1948; 1976, 1977, 1980; Hennig 1966; Zimmermann 1968; Stevens 1980; Cronquist 1988; and many others). There are known many such morphoclines in morphology of vegetative organs (especially in wood anatomy and stomatography), morphology of flowers, palynology, embryology, etc.

The ascertainment of morphoclines raises two questions: 1) Is the given transformation series unidirectional or is it reversible? 2) Which member of the given unidirectional series is the most primitive and which is the most advanced (direction of transformation series or polarity)?

In many series of characters a direction of evolutionary changes is morphogenetically constrained, that is, only one transformation polarity is ontogenetically possible and thus the sequence of stages is determined a priori. For instance, the transformations of tracheids into vessel members, spiral arrangement of floral parts into cyclic arrangement, colpate pollen grains into colporate, apocarpous gynoecia into syncarpous,* superior ovary into inferior, or seeds with endosperm to seeds without endosperm are actually unidirectional and irreversible. The transformation morphoclines are evidently unidirectional also in many reduction series, as, for example, in a progressive reduction of ovules in the order Santalales. Moreover, numerous cases of narrow specialization series (parasitism, xerophilization, etc.) are, as a rule, unidirectional.

*Pseudoapocarpous gynoecia, such as those of some Ochnaceae, are morphologically syncarpous and thus do not contradict the unidirectional sequence from apocarpous gynoecia to syncarpous.

However, in many other morphoclines, directions of evolutionary sequences are less evident or even uncertain. In these cases, the character sequences and trends are frequently determined by statistical methods. The well-known fact that certain characters are statistically associated or correlated led Bailey and Tupper (1918), Frost (1930, 1931), Kribs (1935), Chalk (1937), Sporne (1948, 1976, 1977, 1980), and others to the application of simple statistical techniques for patterns of character distribution. In many cases, for instance in wood anatomy, the study of character correlations is very useful and helps in establishing transformation trends. However, in some other cases it could give doubtful and, even, evidently wrong results (e.g., very few phytomorphologists would agree with Sporne's conclusion of the primitiveness of unisexual flowers).

One can compare a sequence of characters in a given group with patterns of character distribution in related group (or groups), both extinct and extant. It is an "out-group analysis" in a broadened sense. As is widely known, relationships between groups are best expressed between their most archaic members. Therefore, if one extreme of a morphocline resembles a condition found in the less advanced members of related group of the same rank, this extreme is primitive (Maslin 1952). For instance, monocolpate (sulcate) pollen grains of archaic magnoliopsids resemble pollen of such an archaic division of gymnospermous plants as Cycadophyta, which coupled with other data (including palaeobotanical record) confirms the primitiveness of the distal aperture. The out-group analysis, which Stevens (1980) even considers as the most satisfactory method for assigning evolutionary polarity, is certainly one of the major criteria.

Finally, there are also strictly ontogenetic criteria, which assume that the direction of evolutionary sequence of characters corresponds to the sequence of ontogenetic stages. However, as I have already mentioned, a recapitulation may occur only when evolutionary changes take place by addition of end stages (Sewertzoff's anaboly). It is especially true when ontogeny is closed or, according to Tomlinson's terminology, "primordial" (1982, 1984). This kind of ontogenetic processes is characteristic for unitary organisms like vertebrates and for unitary organs of plants like individual leaves, stamens, carpels, pollen grains, seeds etc. Closed ontogenies result in unitary organisms or in single mature modules of modular organisms. All developmental stages of closed ontogenies are so profoundly interconnected, that there is only little possibility for recapitulation. Velenovsky (1910) even concluded that in order to avoid errors and inaccuracies, data on the developmental history of organs ought to be ignored altogether. But this is an extreme view.

The situation is somewhat different in cases of open or serial ("repetitive") ontogeny, which is characterized by developmental processes producing series of homologous* adult structures, as in the succession of leaves along a shoot or the succession of tissues resulted by secondary growth of a stem. Thus, they are both longitudinal and transverse sequences of these serial structures. Therefore, recapitulations, when they occur, are also serial. In contrast to recapitulations manifested in "primordial" developmental processes, serial recapitulations or retentions, as I prefer to name them (Takhtajan 1943), manifest themselves in adult structures of the preceding members of the series. Thus, retentions characterize adult structures rather than transient developmental phases of a given structure. While we observe the stadial recapitulations only in the process of closed ontogeny, the retentions manifest themselves in the adult structures.

A great many examples of retentions may be mentioned.

---

*Homologous entities, including those which occur within individuals, are structurally related and morphologically correspond to each other (see Sattler 1984). According to Van Vallen (1982), homology is resemblance caused by a continuity of information. Serial, repetitive, or modular homology is a special kind of homology characteristic for modular organisms.

They are most obvious in leaf series, especially in plants with strongly modified leaves. For instance, such leaf series are in a number of Australian species of *Acacia* and certain Australian species of *Oxalis*. Their lower leaves have normal blades (retention of ancestral condition), whereas the upper ones are transformed into phyllodes. Despite the sharp difference between the former and the latter, both types are usually connected by intergrades. These retentions of the ancestral, or rather, near-ancestral condition clearly show the direction of leaf morphocline.

Apart from retentions in leaf series, retentions in the structure of vascular system are also widespread. It is well known that in many higher plants certain ancestral characters of the vascular system of the stem are retained in its basal part. For instance, numerous investigations of the stem anatomy of herbaceous flowering plants have led to the conclusion that the basal part of the stem is often similar to the stem structure of woody ancestors. Thus, according to my own observations, the basal part of the stem of *Zygophyllum fabago* is similar to the stem structure of the related woody species such as *Z. atriplicoides*. Serial retentions are observed not only in the "longitudinal" series, i.e., with retentions expressed in an "archaism of bases" (juvenile leaves, early established parts of the vascular system, etc.). There are also "transverse" retentions, which manifest themselves in the secondary growth of stems. For instance, in many flowering plants the first, earliest layers of xylem have a scalariform perforation of vessels which is subsequently supplanted with simply perforated vessels. This transverse serial alteration of two types of perforation reflects the evolutionary sequence. There are a number of other examples of transverse retentions, especially in wood anatomy.

But it is necessary to note that, along with retentions, there are also some cases of "inversions" of the evolutionary sequences. They are found both in leaf series and in the transverse series of anatomical characters. Just as the lowermost leaves may sometimes be more highly specialized than the succeeding ones, so also in the structure of xylem, the early layers may be more highly specialized than the later layers. For instance, elimination of rays in secondary xylem of shrubs disappears in the inner part of the xylem, although they are still present in late xylem. Consequently, inner (earlier) xylem is more advanced than the late xylem, thus, the evolutionary sequence of characters is reversed. This reverse is explained by Barghoorn (1941) purely in terms of phylombryogenesis, for he links it with an accelerating "modification of ontogeny," which begins in early stages.

We come, thus, to the conclusion that there is no one absolute criterion for assigning evolutionary polarity. In different cases different criteria are valid.

References

Arber A. 1937. The interpretation of the flower: a study of some aspects of morphological thought. Biol. Rev. 12:157–184.

Arber A. 1950. The natural philosophy of plant form. Cambridge.

Bailey I. W. and W. W. Tupper. 1918. Size variations in tracheary cells. I. A comparison between the secondary xylem of vascular cryptogams, gymnosperms and angiosperms. Proc. Amer. Acad. Arts and Sci. 54:149–204.

Barghoorn E. S. 1941. The ontogenetic development and phylogenetic specialization of rays in the xylem of dicotyledons. III. The elimination of rays. Bull. Torrey Bot. Club 68:317–325.

Bessey C. E. 1915. The phylogenetic taxonomy of flowering plants. Ann. Missouri Bot. Gard. 2:109–164.

Cain A. J. 1982. On homology and convergence. In K. A. Joysey and A. E. Friday, eds., Problems in phylogenetic reconstruction, pp. 1–19. London.

Chalk L. 1937. The phylogenetic value of certain anatomical features of dicotyledonous woods. Ann. Bot. London n.s. 1:408–428.

Corner E. J. H. 1976. The seeds of dicotyledons. I, II. Cambridge.

Cronquist A. 1988. The evolutionary classification of flowering plants. New York.

Darwin F. and A. C. Seward, eds. 1903. More letters of Charles Darwin. Vol. 1.

Davis P. H. and V. H. Heywood. 1963. Principles of angiosperm taxonomy. Edinburgh and London.

De Beer G. R. 1958. Embryos and ancestors. 3d ed. Oxford.

Engler A. 1892. Syllabus der Vorlesungen über Speciell und Medizinisch-pharmaceutische Botanik. Berlin.

Franz V. 1927. Ontogenie und Phylogenie: das sogenannte biogenetische Grundgesetz und biometabollischen Modi. Abh. Theorie Org. Ent., no. 3. Berlin.

Friday A. E. 1982. Parsimony, simplicity, and what actually happened. Zool. J.Linn. Soc. 74:329–335.

Friis E. M. and W. L. Crepet. 1987. Time of appearance of floral features, In E. M. Friis, W. G. Chaloner, and P. R. Crane, eds., The origins of angiosperms and their biological consequences, pp. 145–179. Cambridge.

Frost F. H. 1930, 1931. Specialization in secondary xylem of dicotyledons. I. Origin of vessels. II. The evolution of the end wall of the vessel segment. III. Specialization of the lateral wall of the vessel segment. Bot. Gaz. 89:67–94, 90:198–212, 91:88–96.

Garstang W. 1922. The theory of recapitulation. A critical restatement of the biogenetic law. J. Linn. Soc., Zool. 35:81–101.

Gould S. J. 1977. Ontogeny and phylogeny. Cambridge (Mass.) and London.

Hallier H. 1912. L'origine et le système phylétique des angiospermes exposés à l'aide de leur arbre généalogique. Arch. Néerl. Sci. Exactes et Nat. Serie 3b (Sci. Nat) 1:146–234.

Hardy A. C. 1954. Escape from specialization. In J. Huxley and A. C. Hardy, eds., Evolution as a process, pp.122–142. London.

Hennig W. 1966. Phylogenetic systematics. Urbana.

Huxley J. S. 1942. Evolution: The modern synthesis. London.

Huxley J. S. 1954. The evolutionary process. In J. S. Huxley and A. C. Hardy, eds., Evolution as a process, pp. 1–23. London.

Kribs D. A. 1935. Salient lines of structural specialization in wood rays of dicotyledons. Bot. Gaz. 96:547–557.

Koltzoff N. K. 1936. The organization of the cell. Moscow and Leningrad. (In Russian.)

Maslin P. P. 1952. Morphological criteria of phylogenetic relationships. Syst. Zool. 1:49–70.

Popper K. R. 1965. Conjectures and refutations. 3d ed. New York.

18   *Introduction*

Popper K. R. 1975. Objective knowledge. An evolutionary approach. Oxford.

Remane A. 1956. Die Grundlagen des naturlichen Systems, der vergleichenden Anatomie und der Phylogenetik. 2d ed. Leipzig.

Rensch B. 1959. Evolution above the species level. New York.

Ridley M. 1986. Evolution and classification. The reformation of cladism. London and New York.

Sattler R. 1984. Homology—a continuing challenge. Syst. Bot. 9(4):382–394.

Sewertzoff A. N. 1927. Uber die Beziehungen zwischen der Ontogenese und der Phylogenese der Tiere. Jena Z. Naturwiss. 56 (o.s. 63):51–180.

Sewertzoff A. N. 1931. Morphologische Gesetzmassigkeiten der Evolution. Jena.

Sewertzoff A. N. 1939. Morphological laws of evolution. Moscow and Leningrad.(In Russian.)

Sporne K. R. 1948. Correlation and classification in dicotyledons. Proc. Linn. Soc., London 160:40–47.

Sporne K. R. 1976. Character correlations among angiosperms and the importance of fossil evidence in assessing their significance. In C. B. Peck (ed.),Origin and early evolution of angiosperms, pp. 312–329. New York.

Sporne K. R. 1977. Some problems associated with character correlations. Plant Syst. Evol., Suppl. 1:33–51.

Sporne K. R. 1980. A reinvestigation of character correlations among dicotyledons. New Phytol. 85:419–449.

Sprague T. A. 1925. The classification of dicotyledons. I. General principles. II. Evolutionary progressions. J. Bot. (London) 63:9–13, 105–113.

Stebbins G. L. 1974. Flowering plants. Evolution above the species level. Cambridge, Mass.

Stevens P. F. 1980. Evolutionary polarity of character states. Ann. Rev. Ecol. Syst. 11:333–358.

Takhtajan A. 1943. Correlations of ontogeny and phylogeny in the higher plants. Trans. Erevan State Univ. 22:71–176. (In Russian with English and Armenian summaries).

Takhtajan A. 1947. On principles, methods and symbols of phylogenetic construction in botany. Bull. Moscow Soc. Nat., Biology 52(5):95–120.

Takhtajan A. 1972. Patterns of ontogenetic alterations in the evolution of higher plants. Phytomorphology 22:164–170.

Takhtajan A. 1976. Neoteny and the origin of flowering plants. In C.

B. Beck, ed., Origin and early evolution of angiosperms, pp. 207–209. New York and London.

Takhtajan A. 1983. Macroevolutionary processes in the history of plant world. Bot. Zhurn. (Leningrad) 68(12):1593–1603. (In Russian with English summary.)

Taylor D. W. and L. J. Hickey. 1990. An Aptian plant with attached leaves and flowers: implications for angiosperm origin. Science 247:702–704.

Tomlinson P. B. 1982. Chance and design in the construction of plants. In R. Sattler, ed., Axioms and principles of plant construction, pp. 162–183.Acta Biotheoretica, vol. 31A. The Hague.

Tomlinson P. B. 1984. Homology: an empirical view. Syst. Bot. 9(4):374–811.

Van Valen L. M. 1982. Homology and causes. J. Morph. 173:305–312.

Vassilczenko I. T. 1965. Neotenous alterations in plants. Moscow and Leningrad. (In Russian.)

Velenovsky J. 1905–1914. Vergleichende Morphologie der Pflanzen. I-IV. Prague.

Wardlaw C. W. 1952. Morphogenesis in plants. London.

Zimmermann W. 1968. Methoden der Evolutionswissenschaft-Phylogenetik. In G. Heberer, ed., Die Evolution der Organismen, pp. 61–160. Stuttgart.

1

Evolutionary Trends in Vegetative Organs

In vegetative characters there are many easily reversible characters, such as growth habit, arrangement, size and form of leaves, but there are also many trends which either can be reversible with great difficulty or are completely irreversible. In general, vegetative organs are characterized by more reversibility than reproductive organs. However, even the most reversible characters usually reveal more or less definite evolutionary trends.

1.1. Growth Habit

The most archaic magnoliophytes are woody plants, trees, or shrubs. The herbaceous habit is always secondary (Hallier 1901, 1905, 1912, Sinnott and Bailey 1914, Jeffrey 1917, and many subsequent authors). The evolution of flowering plants most probably begins with small, relatively weakly branched trees or shrubs. According to Hallier (1912), the early flowering plants were small trees with a weak crown of relatively few thick branches. Stebbins (1974), on the other hand, visualizes the earliest flowering plants as low-growing shrubby plants, having a continuous ring of secondary vascular tissue, able to sprout from the root crown, and no single well-developed trunk. Amongst the living archaic magnoliophytes there are both trees (the majority) and shrubs (*Eupomatia laurina*, for example, is a shrubby plant with several trunks). It is difficult to say with any certainty whether the earliest flowering plants were small trees or *Eupomatia*-like shrubs. We can only say that they were small woody plants, which occupied only a modest and insignificant position in the Early Cretaceous vegetation. Big stately trees of tropical rain forest are derived, having originated from ancestral, small woody magnoliophytes. Trees with numerous slender branches evolved from sparingly branched trees. Deciduous woody plants evolved from evergreen ones.

The derived character of the herbaceous type of stem in the flowering plants is proved by numerous facts both from phylogenetic systematics and morphology. Herbs are completely absent among Magnoliales, Annonales, Winterales, and Trochodendrales and rare among Laurales, but are numerous and frequently predominate among the more advanced orders. The comparison of the woody and the herbaceous forms within individual orders, families, and genera leads to a similar conclusion. A clearly expressed correlation is observed between the extent of herbaceous nature and the level of specialization of the flower and the conducting system of the axial organs. Thus, for example, almost all the herbaceous dicotyledons—with a few exceptions—have vessels with a simple perforation, while the vessels of the related woody forms may have a scalariform perforation. Generally, the herbs are as a rule more advanced than the related woody forms in their structure. In this connection, it is of interest to indicate that the types of the female gametophyte deviating from the normal type are found almost exclusively in the herbs (Ishikawa 1918).

Numerous anatomical data show that the lower part of many herbaceous magnoliophytes resembles in structure the young branches of the woody plants (Eames 1911; Sinnot and Bailey 1914, 1915; Takhtajan 1948, etc.). As the lower part of an herbaceous stem is marked by more primitive traits of the structure than the upper, we are able to follow the chain of changes which led to the herbaceous type through an investigation of the stem from the base to the top. On the other hand, the young shoots of the woody plants have a structure close to that of the herbaceous type and so we can approach the understanding of the origin of the herbs by investigating such shoots and comparing them with older branches. A comparative study of the stems of closely related woody and herbaceous forms led Sinnot and Bailey (1914, 1915) to conclude that the herbaceous stem is essentially the first growth ring of the woody ancestors but with a reduced layer of the secondary wood. The main factor in the origin of the herbaceous stems and roots was reduction in the quantity of the secondary wood due to a decrease in the cambial activity. In this process, a considerable role is also played by an increasing parenchymatization, which occurs mainly due to a widening of the rays.

Thus, the evolutionary transformation of the woody forms into the herbaceous is characterized by a gradual weakening and finally a cessation of the activity of the cambium. Basically, this means a gradual hereditary fixation of the "herbaceous" structures of the woody stem (and root), a fixation accompanied by a greater or lesser modification of the original type. Therefore, the herb may be considered as a fixed juvenile phase of the tree (Takhtajan 1943, 1948, 1959), on which point both Agnes Arber (1950:108) and Golubev (1959) are in agreement. The herbs originated from the trees through a neoteny. This process of neotenic transformation may be traced particularly well in the genera having both herbaceous and woody representatives. One of the best examples is furnished by the genus *Paeonia* (see Takhtajan 1948).

The herbaceous habit originated independently in various

ways along different phyletic lines and at different evolutionary grades. The first herbs had probably originated already at the dawn of the angiosperm evolution, but with the pasage of time the origin of herbs occurred more quickly and on a wider scale. The evolutionary transformation of the woody forms into the herbaceous occurred under the most diverse climatic, edaphic, and biotic conditions. An infinite number of perennial herbaceous forms originated in the moist tropical regions (Bews 1927) owing to the adaptation of climbing, and especially the epiphytic mode of life, as well as due to saprophytism, parasitism, and hydrophilous evolution. Many of these forms originated in the temperate forests. As a result of parasitism, there arose such parasitic forms as *Cassytha* (Lauraceae) or the families Rafflesiaceae, Hydnoraceae, and Balanophoraceae. The peculiar herbaceous groups united in the order Nymphaeales as well as the genus *Nelumbo,* isolated in the system to constitute the distinct order Nelumbonales, probably arose already at one of the early stages of the magnoliophyte evolution due to adaption to the aquatic mode of life. The archaic monocotyledons, which probably sprang up from the remote ancestors of the present day Nymphaeales, are also the product of the hydrophilous evolution under the tropical climatic conditions.

The climatic factors—the cold climate of the high altitudes and polar regions and the climate of the arid regions—had an enormous importance in the origin of the herbs. The adaptation of the magnoliophytes to the conditions of high altitude and polar climate was one of the most important prerequisites of the development of herbs. Gradually advancing to these zones, the woody forms were reduced more and more and transformed into shrubs and perennial herbs. Both of these are well adapted to the cold climate of the arctic and alpine tundras. In the arid regions, the herbaceous forms predominate in the flora so far as the number of species, and more so of the individuals, is concerned. As opposed to the polar regions and high mountains, in the arid countries there is a mass development of the annual forms. The flora of certain types of deserts mainly consists of annual herbs, capable of using the very short wet season with maximum intensity.

Compared with the woody forms, the herbs are more progressive and more plastic in the evolutionary context. The reproductive phase starts early in the herbs and with a minimum expenditure of the construction of the vegetative organs, whereas seed production attains the maximum as compared with the vegetative mass. The herbs consequently have higher reproductive capacity, are more "high-yielding" than the trees and the shrubs. Besides, it is quite evident that the dispersal of the herbaceous species takes place much more quickly than that of the trees. A quicker succession of generations than that of the trees ensures a higher tempo in the evolution of the herbs. The rates of evolution of herbaceous magnoliophytes increase noticeably as compared to trees (Eames 1911). By virtue of these peculiarities, the herbs—especially the annual ones—quickly spread over the earth, attained a very high diversity in forms, became adapted to all possible environmental conditions, and started playing a major role in the vegetation. The development of herbaceous flowering plants had an exceptional significance in the evolution of the animal world, particularly in the evolution of the herbivorous mammals and terrestrial birds.

The evolutionary trend from woody flowering plants to herbs was not irreversible. In some phyletically distant taxa of magnoliophytes, the reverse process of transformation of herbaceous plants into arborescent forms took place, for example, in Ranunculaceae, Berberidaceae, Papaveraceae, Phytolaccaceae, Nyctaginaceae, Amaranthaceae, Chenopodiaceae, Polygonaceae, Cucurbitaceae, Lobeliaceae, Asteraceae, and many liliopsids. But, usually, these secondary arborescent plants, especially arborescent liliopsids, strikingly differ from the primary woody plants. As Stebbins (1974:150) aptly remarks, "Palms and bamboos are as different from primitive preangiospermous shrubs and trees as whales and seals are from fishes."

Apart from the evolution of stem from the woody type to the herbaceous, the stem evolves in many groups along the line of narrower adaptations. Thus the prostrate forms and various lianes as well as numerous epiphytes so typical of certain kinds of wet tropical and subtropical forests, arose out of the upright forms. Besides, the saprophytes as well as the semiparasites and parasites sprang up from the green autotrophic forms. In the course of the hydrophilous evolution, quite a number of aquatic magnoliophytes originated from the land forms. Lastly, during the geophilous evolution, arose the life forms with various kinds of subterranean organs, serving for hibernation, for depositing the reserve nutritive substances, as well as for vegetative propagation.

1.2. Branching

There are two main morphological types of branching in flowering plants—monopodial and sympodial. Both these types are met in many families and even within one and the same genus and change from one to the other with great ease. This makes the determination of the main direction of evolution of the branching in flowering plants somewhat difficult. The study of the most archaic extant magnoliophytes indicates that perhaps the original type has a combination of monopodial and sympodial branching—well expressed, for example, in *Magnolia*. The vegetative branches of *Magnolia* are monopodial, but the short branches carrying the terminal flowers develop in a strictly sympodial manner, and the apparently simple axis of such a branch is in fact a sympode of a certain number of shoots of an ascending series. The sympodial nature of a reproductive branch is determined by the fact that each of the component axes ends in a terminal flower, arresting its subsequent development. So the sympodial nature here is primary and not secondary as in the evolution of the vegetative branches. Monopodial branching is characteristic of many trees of the humid subtropical and particularly the humid tropical forest (Serebryakov 1955:75). This is explained by the fact that the conditions of humid tropical and subtropical climates help in prolonged preservation of the terminal meristems of the stems so that the growth of the vegetative shoot occurs all the time through a continuously operating apical meristem, which leads to a vigorous development of the main axis and to a greater or lesser suppression of the lateral shoots. But in the extratropical regions as well as in the mountains of tropics and under the conditions of a dry tropical climate, the sympodial branching arises out of monopodial (Takhtajan 1948, 1964; Serebryakov 1955). The growth of the annual shoots ends in the disappearance of their terminal bud, which inevitably leads to the development of a large number of lateral buds and the formation of a larger number of lateral shoots. The main axis ceases to hinder the development of the lateral shoots, the intensity of branching is amplified, and the crown becomes denser. The process of the origin of sympodial branching out of the monopodial type is realized in the most diverse phyletic lines and at various levels of specialization. Sympodial branching is very widespread in the herbaceous angiosperms. It is observed in almost all monocotyledons, where it is a direct result of the reduction of the cambium (Holttum 1955), and quite typical of the herbaceous dicotyledons as well. The biological advantages of sympodial branching is emphasized by Zhukovsky (1964:125), who thinks that the successive dying off of the terminal buds should be considered as a very useful adaptation. According to Serebryakov, sympodial renewal was in addition a vigorous tool for intensifying vegetative reproduction (1952:278). Lastly, in his opinion, the dying off of the shoot apex or the terminal buds under sympodial growth provides for an earlier "maturing" of the shoots, their transition to the state of dormancy, and an intensification of the hardiness of the trees and shrubs.

1.3. Evolutionary Trends in Leaves

In the infinite range of diversity of form, structure, and size of leaves, the flowering plants surpass all other groups of seed plants. The extraordinary leaf polymorphism is sometimes found within one family (e.g., Araceae or Arecaceae) or even in one genus (e.g., *Acer* and *Quercus*). This morphological diversity of leaf architecture is, however, brought about by more or less simple mechanisms of differential or allometric growth, which is the most typical mode of deviation of ontogeny from its previous course. The differential growth determines the form of both the organism as a whole and its separate parts—organs, tissues, cells, and organelles. The morphological changes determined by the differential growth can be described mathematically using method of Cartesian coordinates (see D'Arcy Thompson 1917, 1942; and Wardlaw 1952, 1965). Thus, if the shape of a leaf is described on grid paper, its length and width being treated as functions of the X and Y axes, and if, now, the system of coordinates is altered or deformed, a new transformed system of coordinates will be obtained and the inscribed figure will become deformed in a manner which precisely follows the deformation of the coordinate system (Wardlaw 1952:334). In other words, the inscribed leaf form will be transformed into a new but related form, which may more or less correspond to existing ones. The evolutionary changes of the leaf forms may be described as continuous deformations of Cartesian coordinates, that is, as topological transformations. Exceptions are provided only by those cases, where the morphological changes of leaves necessitate some ruptures as in the origin of the perforated leaves of some Araceae, for example, in some species of *Monstera*.

Unfortunately, we do not find any indisputably ancestral primitive leaf type among the extant flowering plants, not even among the most archaic ones. They are absent also among the earliest known fossil forms. It is therefore very difficult to postulate the initial type of the angiosperm leaf.

1.3.1. Evolutionary Trends in Leaf Form

Some authors, including Corner (1949), consider the compound leaf to be the primitive in the flowering plants. But the comparative morphological data led a majority of authors to the conclusion that compound leaves originated from simple, entire, or lobed leaves. But already Hallier (1912:149) had suggested the evergreen, coriaceous, simple, entire, and pinnately veined leaves characteristic for the Annonales (*sensu* Hallier). Sinnott and Bailey (1914, 1915) also concluded that the primitive leaf type is a simple leaf, but according to them, it is a three-lobed leaf with palmate venation rather than an entire leaf with pinnate venation. Later, on the basis of comparative morphology of the leaves of dicotyledons, Parkin (1953:84) came to the conclusion that the simple oval-shaped leaf with a pinnate venation might have been a possible initial leaf type of flowering plants from which other types could be derived. "The first change could be the broadening of the lower part of the lamina to produce an ovate leaf with perhaps also a cordate base and the same time a change in the venation from pinnate to palmate. Then follows the lobing of the lamina. A three-lobed lamina would appear often to precede a five-lobed one. By the deepening of the lobing to the base a palmate compound leaf would be reached. By the interpolation of a rhachis this would become ultimately a pinnate compound one" (Parkin, ibid.). To illustrate these evolutionary trends, Parkin cites the genera *Acer* and

Rubus. I also came to similar conslusions regarding the primitiveness of the simple entire pinnately-nerved type (Takhtajan 1954, 1959), though earlier I shared the views of Sinnott and Bailey. The fact that the most archaic extant magnoliophytes, such as Degeneriaceae, Himantandraceae, Magnoliaceae, Eupomatiaceae, Winteraceae, Illiciaceae, Schisandraceae, Annonaceae, and others, have simple—usually entire—leaves with a pinnate venation was mainly responsible for leading to this conclusion. The lobed leaves are found very rarely among the most archaic flowering plants and only in relatively advanced taxa as the genus *Liriodendron* and certain species of *Magnolia.* The primitiveness of the simple and entire, pinnately veined leaf type is accepted also by Eames (1961), Cronquist (1968, 1988), Hickey (1971), Stebbins (1974), and others.

The simple and entire, pinnately veined leaf gave rise to pinnately lobed and pinnatifed leaves with pinnate venation and palmately lobed and palmatifid leaves with palmate venation. In many groups, from the pinnatifid leaves originated pinnately compound leaves, and from the palmatifid leaves arose palmately compound leaves (figure 1). These trends in leaf evolution are reversible. Thus, in many groups, the compound leaves give rise to unifoliolate compound leaves due to a reduction in the number of leaflets, as in the leaves of *Berberis* and *Citrus.*

1.3.2. Evolutionary Trends in Leaf Venation

Leaf venation is an important taxonomic character which received more detailed study by morphologists and paleobotanists than by taxonomists dealing with the living plants. Rudimentary classifications of the venation types one can find in some botanical textbooks written by great botanists of the last century, including A. P. de Candolle and J. Lindley. Much more detailed classification was made by an Austrian botanist Constantin von Ettingshausen (1861) and later by another Austrian

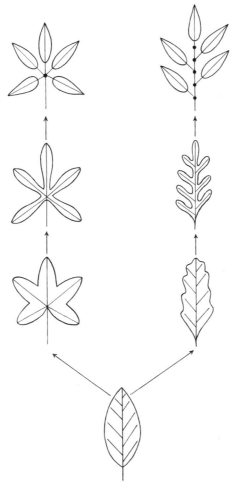

Figure 1. Main evolutionary series of leaf types from the simple leaf with pinnate venation (bottom) to the palmatilobed, palmaticleft, and palmately compound leaves (left series) and pinnatilobed, pinnaticleft, and pinnately compound leaves (right series).

botanist Anton Kerner von Marilaun (1887). Unfortunately, both Ettingshausen and Kerner von Marilaun based their terms on Greek roots, "which results in an ungainly series of polysyllables which do not harmonize with the Latin terminology used in taxonomic descriptions" (Melville 1976:549). In the twentieth century, venation patterns have been studied by a number of botanists, mainly morphologists and paleobotanists (Troll 1939; Takhtajan 1948, 1959, 1964; Foster 1950, 1961; Mouton 1970; Hickey 1973, 1979; Foster and Gifford 1974; Mädler 1975; Hickey and Wolfe 1975; Melville 1976; Spicer 1986).

It is plausible to recognize three major types of leaf (and leaflet) venation—pinnate, palmate, and striate, which in their turn are subdivided into subtypes and varieties.

They are the following:

I. Pinnate venation. With a single prominent median primary vein (midvein or midrib) extending to the leaf apex, along which straight or arching secondary veins are arranged.
  A. Rectipinnate (Melville 1976; craspedodroma— Ettingshausen 1861). Secondary veins running straight or nearly so and terminating at marginal teeth, sometimes even projecting out.
    1. Simple rectipinnate (Melville 1976; craspedodrome simple—Mouton 1970). Veins terminating at marginal teeth without branching. Examples: *Alnus glutinosa, Carpinus betulus, Castanea sativa, Fagus orientalis, Quercus castaneifolia, Dillenia indica, Tetracera alnifolia, Ulmus glabra, Callicoma serratifolia.*
    2. Compound rectipinnate (Melville 1976; craspedodromecomposée—Mouton 1970). Veins branching near the margin to supply several teeth. Examples: *Euptelea polyandra, Corylopsis sinensis, Betula medwedewii, Corylus avellana, Actinidia chinensis, Davidia involucrata, Clematoclethra lasioclada, Viburnum lantana, V. dilatatum.*
  B. Curvipinnate (Melville 1976). Secondary veins curving gradually towards the leaf margin and not supply-

ing a marginal tooth directly or only partly supplying marginal tooth.

1. Simple curvipinnate (Melville 1976; camptodrome —Kerner von Marilaun 1887; eucamptodromous —Hickey 1973). Secondary veins upturned and gradually diminishing apically inside the margin, connected to the superadjacent secondaries by a series of cross veins without forming prominent marginal loops. Examples: *Bridelia ferruginea, Laportea canadensis, Sageretia hamosa, Rhamnus saxatilis, Cornus mas, Mussaenda elegans.*

2. Looped (Stearn 1966) or coarcuate (Melville 1976) (brochidodrome—Kerner von Marilaun 1887; brochidodrome—Mouton 1970). Secondary veins run outwards joining together in a series of arches.

    a) Semilooped (semicraspedodromous—Hickey 1973). Secondary veins branching just with the margin, one or more of the branches terminating at the margin, the others joining the superadjacent secondary and forming loops. Examples: *Kadsura japonica, Chloranthus japonicus, Trochodendron aralioides, Osmanthus fragrans, Saurauia fasciculata, Azara petiolaris, Deutzia discolor, Laurocerasus officinalis, Cerasus avium, Aucuba japonica.*

    b) Simple looped ("festooned brochidodromous" *sensu* Hickey and Wolfe 1975). Secondary veins more or less irregularly branching form several orders of loops gradually diminishing towards the leaf margin; loops are of more or less unequal size and irregular shape. Examples: *Degeneria vitiensis, Eupomatia laurina, Tasmannia piperita, Zygogynum pancheri, Austrobaileya maculata, Illicium anisatum, Laurus nobilis, Tern-*

stroemia tepazapote, Rhododendron ponticum, Syringa amurensis.

c) Multiarched (brochidodrome arche—Mouton 1970; multiarcuate—Melville 1976). Secondary veins forming more or less strong coarcuate inframarginal vein and breaking up into a series of small arching loops forming a zone between the inframarginal vein and the leaf margin. Examples: many Annonaceae, Clusiaceae, Rutaceae, many species of *Ficus, Napoleonaea leonensis, Acridocarpus longifolius.*

d) Paxillate (Melville 1976); brochidodroma—Ettingshausen 1861, brochidodrome marginale—Mouton 1970). Secondary veins numerous, closely parallel to one another and more or less straight, except near margin where they curve more or less abruptly into a submarginal vein, generally making angles of 60 to 90 degrees to the midvein. Examples: *Calophyllum inophyllum, Ficus elastica, F. venosa, Myrcia multiflora, Eugenia corymbosa, Periploca graeca, Allemanda verticillata, Plumeria alba.*

3. Reticulipinnate (dictyodroma—Ettingshausen 1861; reticulidromous—Hickey 1973). Secondary veins losing their identity toward the leaf margin by repeated branching into a vein network. Examples: *Berberis circumserrata, Dendromecon rigida, Rhododendron ungernii.*

C. Palmate-pinnate (Melville 1976). Intermediate between pinnate and palmate, with the distal part of the leaf pinnate and a basal or suprabasal pair of pinnated major veins extending for one-third to two-thirds of the length of the lamina. Examples: *Tetracentron sinense, Tilia* spp., *Acer tataricum, Apeiba tibourbouii,*

Grewia spp., *Thespesia populnea, Erythropalum scandens, Lonicera glabra.*

These basic types of pinnate venation are linked between themselves by many intermediate forms.

II. Palmate venation. Three or more relatively equal primary veins diverge from the leaf base or some distance above the leaf base.

   A. Rectipalmate (Melville 1976); actinodroma marginalis —Ettingshausen 1861). Three or more primary veins diverging radially from a single point at the lamina base or some distance above the base extend more or less straight to the leaf margin. Examples: *Circaeaster agrestis, Kingdonia uniflora, Liquidambar styraciflua,* certain species of *Acer* (including *A. palmatum* and *A. platanoides*), *Gunnera chilensis.*

   B. Reticulipalmate (actinodroma retiformis—Ettingshausen 1861, reticulate-actinodromous—Hickey 1973). Primary veins (except the median one) not reaching the margin and by repeated branching and anastomosing give rise to a vein network. Examples: *Asarum europaeum, Aristolochia manshuriensis, Cercidiphyllum japonicum, Triumfetta* spp., *Tilia mexicana, Cercis siliquastrum, Hedera canariensis.*

   C. Pedate (Melville 1976; pédalée—Mouton 1970; palinactinodromous—Hickey 1973). Leaf palmatilobed or palmatifid, with the upper lobes supplied by primary veins but lower lobes on either side supplied not by primary veins, but by secondary rectipinnate laterals of the lower primaries. Examples: *Platanus occidentalis, Bryonia alba, Curcurbita pepo, Lasia aculeata.*

   D. Curvipalmate (convergate or curvipalmate—Melville 1976, pro parte; acrodroma—Ettingshausen 1861, pro parte). Three or more primary veins or their branches, originating at, or close to, a single point and running in recurved and more or less converging arches toward the leaf apex. Examples: *Saururus cernuus,* certain species of *Peperomia, Cinnamomum zeylanicum, Pilea smilacifolia,* certain Melastomataceae, *Plantago major, Melaleuca leucadendron, Viburnum cinnamomifolium, Strychnos* spp.

III. Striate venation (Troll 1939; Foster and Gifford 1974) and derived types. Three or more bundles enter separately the lamina giving rise to three or more separate primary veins which run toward the apex of the lamina and gradually converge. Venation is almost always closed, without free ends.

   A. Arcuate-striate (Troll 1939; Foster and Gifford 1974; campylodroma—Ettingshausen 1861). Several primary veins running in recurved arches toward the leaf apex and gradually join the adjacent inner primaries as they converge in the upper region of the lamina. Primary veins are usually connected by more or less transverse crosspieces or commissural veinlets. Examples: *Hydrocharis morsus-ranae, Hydrocleys nymphoides, Alisma plantago-aquatica, Potamogeton natans, Veratrum album, Hosta japonica, Maianthemum bifolium, Smilax aspera, Dioscorea sativa, Tacca cristata.*

   B. Pedate-striate. Inner primaries on both sides of the lamina turn inward and converge separately. Examples: *Sagittaria sagittifolia, Avetra sempervirens.*

   C. Palmate-striate. Several primary veins diverging radially from a single point at the lamina base extend more or less straight to the ends of leaf lobes. Examples: *Dioscorea brachybotrya, Anthurium macrolobium.*

   D. Pinnate-striate (Troll 1939; Foster and Gifford 1974). With many primary veins making angles of 60 to 90 degrees to the median veins and joining at equal intervals to the upper adjacent primaries. Examples: *Pentastemona sumatrana, Tacca plantaginea, Canna indica, Heliconia cannoidea, Lysichiton camtschatcense, Calla palustris, Anthurium elegans.*

E. Curvimarginal (parallélodrome transverse—Mouton 1970; curvi-paxillate—Melville 1976). With numerous closely parallel secondary veins, arching to nearly straight, except near the margin where they curve more or less abruptly into a marginal vein. Examples: *Strelitzia reginae, Ravenala madagascariensis, Musa* spp., *Heliconia* spp., *Calathea sebrina.*

F. Lyrate (Melville 1976). With numerous parallel forwardly directed (oblique) secondary veins making angles of 10 to 30 degrees with the midvein consisting of a few to many closely aggregated bundles. Examples: *Dracaena* spp., *Cordyline* spp., *Hanguana malayana.*

G. Longitudinally striate (Troll 1939; Foster and Gifford 1974; parallelodroma—Ettingshausen 1861; parallelodrome longitudinale—Mouton 1970; collimate—Melville 1976) or parallel. With many primary veins running longitudinally to the leaf apex. Examples: Hyacinthaceae, Alliaceae, Convallariaceae, Orchidaceae, Juncaceae, Cyperaceae, Poaceae.

The proposed classification is not comprehensive and is not yet completely evolutionary. There are numerous intermediate forms which with equal right could be ascribed to different venation types. There are also many venation forms which only by stretching a point could be put in the procrustean bed of classification. A construction of more extensive, detailed, and evolutionary classification of the venation types demands a broader comparative study of the angiosperm leaves as well as an ontogenetic and functional approach.

In spite of numerous cases of convergent evolution and very many cases of reversals, leaf venation patterns show some definite evolutionary trends.

The leaves of a majority of magnoliopsids and some of liliopsids are characterized by one or other type of pinnate venation. One of the most significant trends in the evolution of pinnate venation is a gradual strengthening of the role of the midrib and the petiole. The strengthening of the midrib is connected with the intensification of its role as the main arterial line, while the strenghtening of the petiole is related mainly to its mechanical functions. A strong development of the midrib and the petiole is especially typical of the evergreen leaves of the trees of tropical rain forests. The leaves of these plants are often large and heavy and therefore have a strong cylindrical petiole. Such petioles as well as strong midribs, are good elastic springs, which resist effectively such dynamic actions as gusts of wind, impact of raindrops at the time of heavy shower. (Razdorsky 1955).

Another evolutionary trend in pinnate venation is the gradual change of the angle of divergence of the secondary veins. In most primitive type of pinnate venation, e. g. , in many Magnoliidae, *Trochodendron,* some Hamamelidaceae, *Corylus, Populus,* many Rosanae, *Stachyurus,* Ulmaceae, Rhamnaceae, Vitaceae, *Viburnum,* the secondaries diverge from the midrib at a very acute angle. But, in more advanced types, the angle formed by the secondaries increases and more and more approaches a right angle, as in *Ficus elastica* and some Apocynaceae and Asclepiadaceae.

The study of evolutionary trends in pinnate venation poses many difficulties. We still do not know with any certainty which type of pinnate venation is the most primitive. Stebbins (1974:331) suggested that the leaves of the original angiosperms were tapered at the base to an indistinct petiole, had a netted venation which lacked free endings, and their primary, secondary, tertiary, and quaternary veins were less differentiated. Besides, paleobotanical data bring some authors to the conclusion that the most primitive type of venation is the "brochidodromous arching" venation with highly irregular size and shape of areas between secondary veins, the irregularly ramifying courses and poor differentiation of the tertiary and higher vein orders, and the frequently poor demarcation of petiole from lamina (Hickey 1971; Hickey and Doyle 1972; Doyle and Hickey 1976). According to Cronquist (1988:177), the best comparison of primitive fossil leaves with modern leaves is those of some members of the Winteraceae, such as *Tasmannia* (*Drimys*) *piperita* and *Zygogynum pancheri* (figure 2). These models of primitive venation, especially that of Cronquist, correspond to the venation type which I call simple looped. This primitive type is characteristic for many Magnoliidae.

One of the main evolutionary trends of looped venation is the origin of semilooped (semicraspedodromous) venation characteristic for many Dilleniidae and Rosidae. The origin of this venation type is connected with the transition from the entire leaf lamina to dentate and serrate one. The culmination of this trend is the origin of rectipinnate (craspedodromous) venation.

Another main trend is the origin of multiarched venation which is charactristic for very many magnoliopsids. Multiarched venation originated as a result of specialization of loops of different branching ranks and the increasing dominance of the inframarginal coarcuate vein.

One of the climax types is Melville's paxillate venation. During the evolution of the pinnate venation the secondary veins often become very tender, stretch along a straight line from the midrib almost to the very margin but at once bend here almost at a right angle and extend to the margin. As a result of the straightening and fusion of the arch segments, is formed an inframarginal vein which appears to be an independent vein.

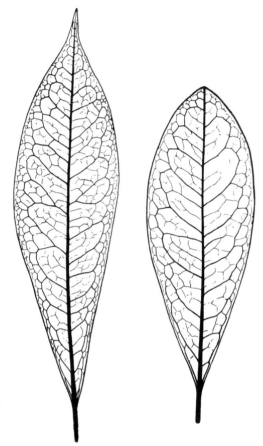

Figure 2. Leaves of *Zygogynum pancheri* (above), and *Tasmannia piperita* (below), showing irregularly pinnate venation characteristic of the Winteraceae (from A. Cronquist 1988).

This very advanced variation of the looped venation is well expressed in *Calophyllum inophyllum* and *Ficus elastica*.

All other types of the pinnate venation most likely also evolved from the simple looped type.

The palmate venation emerged by only a small variation of the primitive pinnate venation (more intense development of the lower secondary veins and weak development of the upper ones). The palmately veined leaves originate with particular ease in the herbs and in the temperate deciduous woody forms, that is, in plants where a strong petiole and midrib are not necessary. The palmate venation also appears in small sessile phyllomes (cotyledons, prophylls, bud scales, bracts, bracteoles, sepals, and petals). The palmate-pinnate venation is one of intermediate forms.

The most primitive type of palmate venation is rectipalmate (actinodromous—*sensu stricto*) venation, characteristic, e.g., for *Acer palmatum*. All other types of the palmate venation, such as reticulipalmate, pedate, and curvipalmate, are seemingly derived. The typical curvipalmate venation, characteristic, e.g., for *Saururus cernuus* is in some respects an intermediate form between typical palmate venation and the striate venation of liliopsids.

The arcuate-striate (campylodromous) venation appears as the most primitive form of striate venation. It characterizes many liliposids, including some relatively archaic forms. But the most characteristic "monocotyledonous" venation is the so-called "parallel" venation. The designation of this type as "parallel" is inaccurate and misleading. Careful examination of typical parallel venation in grasses and other groups "reveals that the main veins do not extend equidistant throughout their course but, on the contrary, converge and progressively anastomise toward the apex of the lamins" (Foster and Gifford 1974:560). Following Troll's term "streifigen Typus Nervatur" (1939:1068), Foster and Gifford prefer to term this type of venation "striate" rather than "parallel" (figure 3). The parallel or longitudinally of striate type of venation originated very early and existed already in the Aptian.

In some cases, from typical striate venation there arose secondary pinnate venation (pinnate-striate, according to Troll and Foster and Gifford), for example, in Zingiberales, some Areca-

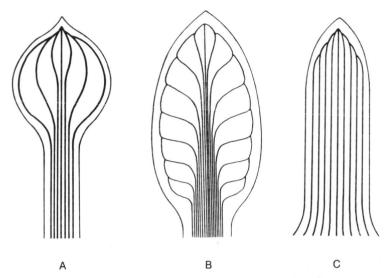

Figure 3. Diagrams showing some types of striate venation. A, arcuate-striate venation pattern; B, pinnate-striate venation pattern; C, longitudinally striate venation pattern (after Troll 1939; from A. S. Foster and E. M. Gifford 1974).

ceae and Araceae, and some other liliopsids. There are many intermediate states between arcuate-striate and pinnate-striate venation. In some cases, as in *Anthurium* spp., they have developed inframarginal vein.

A very special kind of striate venation is a lyrate venation. It probably originated from the longitudinally striate venation.

1.3.3. Evolutionary Trends in the Structure of Minor Veins

For the evolutionary morphology of leaves, also very important is the structure of the minor veins (finest veinlets of the ultimate vein order) which play a major role in transport of water and the products of photosynthesis. They carry out the functions of transportation, the uptake of the products of photosynthesis, and their loading into the phloem. Specialized cells

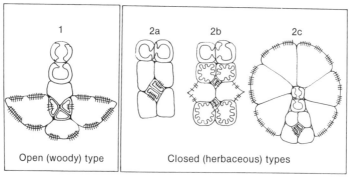

Figure 4. Main structural types of minor veins in dicotyledons. 1, open (woody) type, with numerous plasmodesmal fields between mesophyll and phloem; 2, closed (herbaceous) type, without plasmodesmal contacts to mesophyll. This type includes three subtypes: 2a, with smooth inner surface of cell wall of intermediate cells; 2b, with wall ingrowths in intermediate (transfer) cells; and 2c, with sheath bundles, connected to mesophyll symplastically, but not connected to phloem (from Y. Gamalei 1989).

of the minor veins called intermediary cells mediate between mesophyll and the sieve elements and are involved in the transfer and loading of the photosynthate. Two pathways of this transportation exist—symplastic (from protoplast to protoplast via plasmodesmata) and apoplastic (confined to the cell walls, which constitute the free space together with the intercellular spaces) (see Esau 1977).

According to Gamalei (1988a, 1988b, 1989) there are two structural types of minor veins—open and closed (figure 4). They differ by the structure of the intermediary cells and by the mechanism of phloem loading and sugar transport. The open or arborescent type is characterized by intermediate cells with numerous plasmodesmal fields, by symplastic transport as a main phloem loading mechanism, as well as oligosacharides and other complex sugars as the main phloem transport substances. By contrast, the closed or herbaceous type is characterized by intermediate cells without plasmodesmal connections, as well as by apoplastic loading of sucrose only by membrane proton cotransport. The closed type is divided into three subtypes, differing in the degree of development of the structures used for the sugar uptake from the apoplast.

As Gamalei (1988a) notes, the system of symplastic transport characteristic for the open type is more ancient and is already well developed in brown algae. It occurs mainly in woody plants including such archaic groups as Magnoliales, Illiciales, Annonales, and Laurales, as well as in the Hamamelididae. But it is also found in some advanced woody taxa, such as Ebenales, Myrtales, Elaeagnales, Vitaceae, and Oleaceae as well as in some herbaceous groups (e.g., Cucurbitaceae and Lamiaceae). The closed type is probably of later origin and occurs mainly in herbaceous magnoliopsids (including Caryophyllidae and Asteridae) and almost all of liliopsids.

According to Gamalei (1988b), one can assume that herbaceous forms with open structure of the bundles (many species of Cucurbitaceae, Euphorbiaceae, Onagraceae, Scrophulariaceae, Lamiaceae) evolved relatively recently from the arborescent form and retained the open bundles. On the other hand, woody forms with the closed type of the minor veins one should hypothetically consider as secondary arborescent, derived from the herbaceous ancestors, as—for example—some arborescent Chenopodiaceae.

The direction of the first division of the phloem initial greatly influences the structure of the adult bundles and on the basis of this character it is possible to discriminate bundles of woody and herbaceous plants. The anticlinal division of the phloem initial of the bundles (as distinct from the periclinal one in herbaceous plants) shows the possibility of the formation of cambium, that is, capability of secondary growth. Therefore, according to Gamalei (1988b), this character should have more importance for the analysis of the evolutionary trends of growth forms in different systematic groups, than the adult structure of the minor veins in which the vestigial characters could disappear. It is an important character: the first division of the phloem initial is either anticlinal or periclinal.

1.3.4. Leaf Vernation and Leaf Arrangement

There are more than a dozen manners in which leaves are arranged within the bud (see Lindley 1830; Troll 1939; Davis and Heywood 1963). The most primitive is probably conduplicate vernation when the lamina is folded once adaxially along the midvein so that the two halves of the upper side face one another. Conduplicate vernation is characteristic for some archaic taxa including Degeneriaceae, Magnoliaceae, Eupomataceae, and Winteraceae. All other types, including the circinate (when the leaf is rolled spirally from the apex downwards, as in *Drosophyllum*), are derivative.

The most primitive type of leaf arrangement (phyllotaxy) is alternate arrangement (Hallier 1912 and many others). Both the opposite and verticillate types are derived from the alternate arrangement. But as Cronquist (1988:176) points out, "the origin of opposite leaves from alternate ones is not an immutable trend" and is subject to reversal. As regards verticillate leaves, they are much less reversible.

1.4. Stomatal Apparatus

The stomatal apparatus of flowering plants is characterized by diversity of structure. Stomata may be surrounded either by ordinary epidermal cells (the anomocytic type characteristic of Ranunculaceae, Berberidaceae, Liliaceae, and many other families), or by two or more subsidiary cells morphologically distinct from the other epidermal cells (paracytic, tetracytic, anisocytic, diacytic, actinocytic, and other types)—see Metcalfe and Chalk

(1950); Stebbins and Khush (1961); Esau (1965); Van Cotthem (1970); Patel (1979); Wilkinson (1979); Rasmussen (1981); Inamdar, Mohan, and Subramanian (1986); Baranova (1987a, 1987b, 1987c).

There are two basic types of development of stomata with subsidiary cells—perigenous and mesogenous (Florin 1933). In the first type, the guard cells originate by a single division of the stomatal initial while some of the neighbouring cells become modified as subsidiary cells. In the second type, both the guard cells and subsidiary cells are produced from the same initial. There is also an intermediate mesoperigenous type of stomatal ontogeny (Pant 1965). In the evolution of seed plants the perigenous type preceded the mesogenous type (Florin 1933, 1958), but the flowering plants most probably began with the mesogenous type. This is supported by the occurrence of the mesogenous (and mesoperigenous) type in such archaic families as Degeneriaceae, Himantandraceae, Magnoliaceae, Eupomatiaceae, Annonaceae, Canellaceae, Winteraceae, and Illiciaceae. Moreover, the stomatal apparatus of the mesogenous and mesoperigenous Magnoliidae is of the paracytic type (accompanied on either side by one or more subsidiary cells parallel to the long axis of the pore and guard cells). Mesogenous paracytic stomata are the most primitive and initial type of the magnoliophyte stomatal apparatus (Takhtajan 1966, 1969; Baranova 1972, 1985, 1987a). All other types of stomata, including anomocytic type, which is devoid of subsidiary cells, are derived.

As regards stomatal ontogeny, most of the morphological types of stomatal complexes with subsidiary cells are in fact ontogenetically heterogenous (Baranova 1975, 1987a). According to Rasmussen, "several developmental pathways may result in the formation of indistinguishable mature stomatal complexes and similar sequences of cell divisions may result in quite different structural patterns at maturity" (1981:203). Therefore, as Baranova (1975, 1987a) pointed out, such diversity narrowly limits the taxonomic value of stomatal ontogeny.

Moreover, both the structural and ontogenetic types of stomatal complexes are not evolutionarily irreversible. The reversibility of stomatal complexes and especially of their ontogeny is especially convincing in the case of liliopsids. The liliopsids have both anomocytic stomata and stomata with subsidiary cells. The stomatal complexes with subsidiary cells are usually perigenous. Surprisingly, mesogenous stomata are reported from genera of tribes Cranichideae (N. H. Williams 1979) and Orchideae (Rasmussen 1981), which are obviously of secondary origin.

1.5. Nodal Structure

It is generally agreed that in gymnospermous plants the unilacunar node structure is more primitive, and the multilacunar nodes of cycads and *Gnetum* are derived. But the evolutionary trend in nodal structure of flowering plants is much more debatable. In addition to unilacunar and multilacunar nodal types in flowering plants there is a third type, the trilacunar, unknown in gymnosperms. The presence of three different basic types of nodal structures complicates the situation and makes more difficult the ascertaining of the evolutionary trends in magnoliophytes.

At different times, and by different authors, each of these three types has been accepted as the most primitive and basic nodal structure in flowering plants. The study of all the available data accumulated in literature brought me to the conclusion, that Sinnott's (1914) theory of the primitiveness of the trilacunar type, based on the extensive reconnaissance of 164 families of Magnoliopsida, is nearest to the truth. It also much better corresponds to the widely accepted theory of the primitiveness of the magnolialian stock. The presence of trilacunar nodes in such an archaic family as the Winteraceae, as well as in Himantandraceae, Annonaceae, Canellaceae, Myrisiticaceae, Tetracentraceae, Cercidiphyllaceae, and in the orders Ranunculales, Caryophyllales, Hamamelidales, Dilleniales, and Violales is very suggestive. But some members of the Magnolianae are penta- or multilacunar. Such an extremely archaic genus as *Degeneria* has pentalacunar nodes (Swamy 1949; Benzing 1967) and in the genus *Eupomatia*, which in its vegetative anatomy is one of the most archaic among the vessel-bearing magnoliophytes, the nodes are multilacunar (Eames 1961; Benzing 1967). The nodal structure of the Magnoliaceae is usually also multilacunar (6 to 17 gaps), except in the relatively archaic genus *Michelia*, which is tri-pentalacunar (see Ozenda 1949). This distribution of tri-, penta-, and multilacunar types most probably indicates that tri-and pentalacunar nodes are primitive and multilacunar nodes are derived. But it is much more difficult to decide which of these two types, trilacunar and pentalacunar, is the basic one (figure 5). In my opinion it is quite possible that the earliest magnoliophytes were tri-pentalacunar, like the living genus *Michelia* (Takhtajan 1980).

The unilacunar nodal structure, which Sinnott (1914) considered as having arisen by reduction from the trilacunar, is according to Marsden and Bailey (1955) the most primitive and ancestral nodal type in all seed plants, including angiosperms. They considered the primitive node to be the unilacunar type with two discrete leaf traces. This new concept of nodal evolution was based on the fact that the unilacunar node with two distinct traces is characteristic not only for some ferns and gymnosperms (as was well-known earlier), but also occurs in certain magnoliopsids (Laurales, certain Verbenaceae, Lamiaceae, and Solanaceae). Also, it is repeatedly found in the cotyledonary node of various flowering plants. Bailey (1956) concluded that we could no longer think of the unilacunar node of

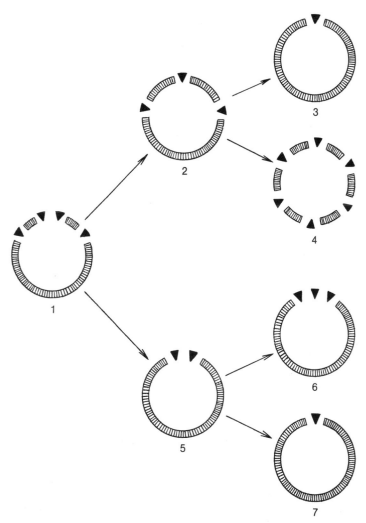

Figure 5. Probable course of evolution of nodal structure in dicotyledons. 1, tri-multilacunar node with a double trace at the median leaf
gap; 2, trilacunar node with three traces; 3, unilacunar node with one
trace; 4, multilacunar node with many traces; 5, unilacunar node with
a double trace; 6, unilacunar node with three traces; 7, unilacunar
node with one trace (from A. Takhtajan 1964.)

dicotyledons as having arisen by reduction from the trilacunar; in his opinion, "during early stages of the evolution and diversification of the dicotyledons, or of their ancestors, certain of the plants developed trilacunar nodes, whereas others retained the primitive unilacunar structure." Canright (1955), Eames (1961), Fahn (1974), and several other anatomists have even more strongly favored the primitiveness of the unilacunar node with two traces, which they consider the basic type in the evolution of angiosperm nodal structure. But there are also objections. Thus Benzing (1967) has pointed out that the occurrence of plants with two-trace unilacunar nodal structure proposed as primitive by Marsden and Bailey (1955) is limited to a few families characterized by mainly decussate phyllotaxy and many specialized floral characters. He also correctly points out that the anatomy of cotyledonary nodes does not necessarily reflect ancestral conditions in the mature stem. "The unique seedling morphology and decussate insertion of the cotyledons make this unlikely," says Benzing. He comes to the conclusion that either the unilacunar node with one trace or the trilacunar node with three traces is more likely to be primitive in the flowering plants than the unilacunar node with two traces. Bierhorst (1971) is also very skeptical about the theory of primitiveness of two-trace unilacunar type.

In my opinion, neither of the two types of unilacunar nodes is primitive and basic in flowering plants (Takhtajan 1980). The unilacunar nodal structure is characteristic mostly for the advanced taxa. In the Magnolianae the unilacunar node is present only in orders Austrobaileyales, Lactoridales, Laurales, and Illiciales, which are considerably more advanced than the Magnoliales. The only unilacunar members of the whole subclass Hamamelididae are *Euptelea* and *Casuarina*. On the other hand, it is significant that the unilacunar node is characteristic for such advanced orders as Ericales, Ebenales, Primulales, Myrtales, Polygalales, Gentianales, Scrophulariales, Lamiales, and Campanulales. Among the sympetalous dicotyledons only some Planta-ginaceae and Asteraceae are exceptions. In some orders, such as Celastrales and Santalales, it is possible to follow the transition from the trilacunar to the unilacunar type, which occurs along with general specialization of the vegetative organs. It is particularly well shown in the family Icacinaceae (see Bailey and Howard 1941). One may see the same evolutionary trend in the series Dilleniales-Theales or Violales-Tamaricales. All these facts lead to the conclusion that the unilacunar type of nodal structure is secondary in flowering plants, having originated from the basic tri-pentalacunar type.

1.6. Evolutionary Trends in Tracheary Elements of Axial Organs

Some of the archaic woody magnoliopsids such as Wintera-ceae, *Amborella, Trochodendron,* and *Tetracentron* have no vessels, and their secondary xylem is almost as primitive as in some cycads.* With the exception of Nymphaeales, all the primarily vesselless dicotyledons are woody plants—trees or rarely shrubs. But Nymphaeales also arose, in the final analysis, from the woody ancestors, though the intermediate forms are completely unknown. In all these vesselless dicotyledons, the water-conducting elements are the tracheids. In the protoxylem these tracheids have annular or spiral thickenings but in the metaxylem these are usually scalariform. In a majority of cases, they are very long (e.g., in *Tetracentron sinense* the average length is about 4. 5 mm). In this case, the tracheids continue to carry out a double function—that of support and water conduction. Vesselless forms exist among the monocotyledons as well. Such, for

---

*In some highly specialized succulents, saprophytes, parasites, and aquatic plants, vessels undergo an evolutionary elimination. Such plants with the reduced xylem tissue are secondary vesselless.

example, are Hydrocharitaceae, Zannichelliaceae, Zosteraceae, Ruppiaceae, etc. (Cheadle 1953). All these vesselless monocotyledons are herbaceous. At least some of them are possibly primarily vesselless. But the absence of vessels in such a family as Lemnaceae is beyond doubt a secondary phenomenon caused by a general degeneration of these plants.

1.6.1. Origin and Evolution of Vessels

Relatively few flowering plants are primitively vesselless. An overwhelming majority of them have vessels, which are the main elements concerned with the conduction of water and substances dissolved in it. Separate cells of the vessels are called the vessel elements or vessel members. They differ from the tracheids in that the tracheids are imperforate cells, whereas the overlapping walls of vessel elements are provided with openings, termed perforations. In the course of evolution, the vessel elements arose from the tracheids. One of the great botanists of the last century, Anton de Bary (1877) first expressed the idea that the scalariform perforation of the vessels originated from transversely elongated bordered pits (scalariform pitting) of the tracheids. Later in the twentieth century it was clearly shown, by a series of anatomical studies, that the tracheids were transformed into vessels due to the disappearance of the pit membranes in the scalariform bordered pitting at the points of contact of the end walls of the adjacent tracheids having extensively overlapping, tapered ends. At the first stages of the origin of the vessel element, the whole process of the evolutionary change consists only in the gradual disappearance of the pit membranes (see Carlquist 1988:58), from which it can be concluded that it is only one step from the scalariformly pitted tracheid to the primitive vessel element. According to Bailey, "this particular phylogenetic sequence clearly is a undirectional and irreversible one, and cannot be read in reverse" (1956:271).

Vessels evolved independently in different divisions of vascular plants and even independently in diverse lineages of Magnoliophyta. According to Bailey (1957:244), the wide range of variation in the reproductive organs of vesselless dicotyledons suggests that vessels may have originated more than once. This idea of Bailey's finds confirmation in the considerable systematic isolation of the vesselless families of the dicotyledons. Though *Trochodendron* and *Tetracentron* are relatively close to each other, they stand quite apart from the Winteraceae and evolved in a completely different direction. If the order Hamamelidales actually sprang up from the ancestors of the present day Trochodendrales, as thought by some, the vessels in it should have originated completely independently from the remaining lineages of magnoliopsids. The vesselless genus *Amborella* stands apart from the Winteraceae as well, the more so from Trochodendrales. It stands much closer to the vascular Monimiaceae than to any of the vesselless groups. But it is still more interesting that the vessels in the monocotyledons originated, in all probability, completely independently of the dicotyledons (Cheadle 1953, Bailey 1944, 1949). According to Bailey, "The independent origins and specializations of vessels in monocotyledons and dicotyledons clearly indicate that if the angiosperms are monophyletic, the monocotyledons must have diverged from the dicotyledons before the acquisition of vessels by their common ancestors" (1944:425). Among the modern dicotyledons, the group closest to the monocotyledons is the vesselless order Nymphaeales. Though the representatives of this order are aquatic plants, the absence of the vessels can hardly be considered as a result of reduction in the present case. An aquatic genus *Nelumbo* and many typical aquatic liliopsids have vessels, e.g., Butomaceae, Alismataceae, Scheuchzeriaceae, Juncaginaceae, Potamogetonaceae, and Pontederiaceae. The vessels could have been preserved also in the roots or the rhizomes of Nymphaeales, the more so as the water lilies are much more archaic than all these families. But vessels are completely absent in Nymphaeales sensu stricto, and the only water-conducting elements of the root and stem in them are very long tracheids of very primitive type. So it seems highly probable that the order Nymphaeales is a primarily vesselless group and arose directly from certain archaic vesselless dicotyledons. The monocotyledons, on the other hand, originated in all probability from the remote ancestors of the present day Nymphaeales. If it is so, then the primitive monocotyledons also should have been vesselless. The results of the comparative anatomical study of the monocotyledons are in full agreement with this conclusion. It is necessary to add that the vesselless monocotyledons, as well as those with primitive vessels only in the roots, belong to so diverse phyletic branches as to compel one to accept the proposition that the vessels arose independently in the monocotyledons several times during their evolution. Cheadle (1953) is also inclined to this view.

Bailey's studies (1944) show that in the dicotyledons vessel elements appeared first in the secondary xylem, then in the late primary xylem (metaxylem), and last in the early primary xylem (protoxylem). In many dicotyledons, the protoxylem may be entirely devoid of vessels. According to Bailey (1944:423), the occurrence of vessels in the first-formed part of the primary xylem is thus an indication of extreme structural specialization. In the monocotyledons, origin and specialization of vessels occured first in the late metaxylem and subsequently worked back into the protoxylem (Cheadle 1943a, 1943b, 1953). Significantly, there are no vessels in the secondary xylem of those monocotyledons which form a well-developed secondary body (Bailey 1944:424). As Cheadle (1943b, 1944) has shown, in the monocotyledons vessels appeared first in the roots and progressed upward in the plant body.

The original vessel types and the lines of their specialization are very similar in dicotyledons (Frost 1930; Bailey 1944) and

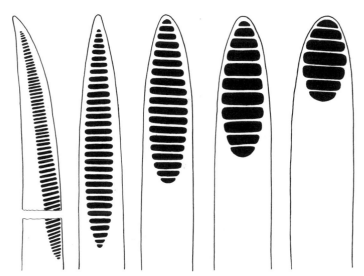

Figure 6. Stages in the evolution of scalariform perforation (from A. Takhtajan 1959).

monocotyledons (Cheadle 1943a, 11943b, 1944; Bailey 1944) (Figures 6 and 7).

The most primitive vessel elements much resemble tracheids in their long fusiform shape, and frequently they are externally almost indistinguishable. They are still comparatively very long* and narrow, thin-walled with an angular cross-section and gradually tapered towards the end, i.e., they do not have the end wall or it is very indistinct (highly oblique). They differ from the tracheids only in the fact that the pit membranes of the scalariform bordered pits disappear at a certain stage of the development. The area of the wall of a vessel element that is perforated is known as the perforation plate. The scalariform perforation plate in the primitive vessel elements consists of numerous (up to 100–150 and more) clearly bordered and narrow slit-like perforations. The lateral walls of the primitive

*But even the most primitive vessel members rarely exceed 1.5 mm in length.

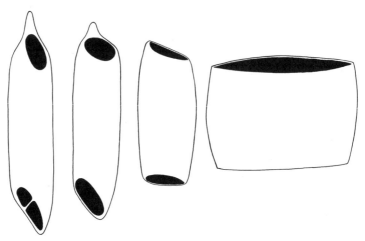

Figure 7. Stages in the evolution of simple perforation (from A. Takhtajan 1959).

vessel elements usually have scalariform pitting. The primitive tracheid-like vessel elements progress gradually and are perfected by specializing more and more along the line of executing the water-conducting function (see especially Frost 1930, 1931; Bailey 1944, 1953, 1957; Carlquist 1975; Esau 1977). In this connection, they shortened progressively, became wider, and developed thicker walls. The cross-section of the vessel elements becomes more and more circular. The scalariform lateral wall pitting of the primitive vessels is replaced by more or less circular bordered pits (figure 8) which are arranged first in horizontal (opposite) rows, and then their arrangement becomes alternate. In the process of evolution of the primitive vessel elements, their end walls became less inclined and finally assumed a transverse position. The structure of the perforation plate also changed in close connection with the variation in the inclination of the end wall. The evolution of the scalariform perforation plate is directed towards the diminution in the number of openings. The initial type is the perforation plate with numerous slitlike openings separated by bars which still preserve their borders,

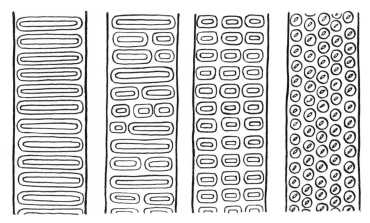

Figure 8. Stages in the evolution of lateral pitting of the vessels from scalariform through intermediate and opposite to the alternate type (from A. Takhtajan 1959).

inherited from the scalariform bordered pits. Along with a decrease in the number of bars, the width of the openings increases, enhancing the rapid movement of water. The successive decrease in the number of the bars is caused by the need to facilitate the free passage of water from cell to cell. The subsequent step is their complete disappearance. Due to the disappearance of all the bars, a simple perforation plate with one large and usually circular perforation is formed. The simple perforation is the most specialized type of opening between the vessel elements (figure 7). Thus, as a result of all these evolutionary changes, the primitive tracheidlike vessel elements are transformed into a highly specialized structure.

The evolutionary trends of increasing specialization of vessels both in dicotyledons and monocotyledons have been established independently of phyletic trends established by systematists on the basis of morphology of reproductive organs. "It is important to bear in mind"—says Bailey "in this connection— that this can be accomplished *entirely independently* of the various systems of classifying the taxa of angiosperms, thus avoiding circular arguments based upon assumptions regarding the primitive or specialized character of various representatives of the angiosperms. In other words, primitive vessels are distinguished from spcialized ones *solely* upon their own structural differences, and *entirely* without reference to the putative primitiveness of the plants in which they occur" (1957:244).

Whereas among the woody dicotyledons one can observe the whole spectrum of the evolutionary specialization of vessels, the herbaceous dicotyledons have, as a rule, mostly vessels with simple perforation and much less frequently with both simple and scalariform perforations. Only few herbaceous dicotyledons have exclusively scalariform perforation plates. They include members of the small families Chloranthaceae, Saururaceae, Paeoniaceae, Penthoraceae, and Pentaphragmataceae and the genus *Acrotrema* (Dilleniaceae). As Cheadle (1953) points out, the perforation plates in these herbaceous dicotyledons are often not at all primitive and have a relatively small number of bars.

1.6.2. Origin and Evolution of Sieve Tubes

Whereas vessel elements arose from tracheids, sieve-tube members originated from the sieve cells of gymnospermous plants. Sieve cells have relatively undifferentiated sieve areas which are rather uniform on all walls and have narrow pores. Sieve-tube elements of flowering plants have generally more differentiated sieve areas—on some walls they have larger pores than those on the other walls. Sieve areas with larger pores— sieve plates—are generally located on the end walls. They have developed a higher degree of spcialization for long-distance assimilate transport in sieve tubes. The activity and the normal functioning of both the sieve cells and sieve elements becomes possible only by the virtue of close physiological interaction with the parenchyma cells adjacent to them. The interaction is facilitated by the albuminous cells in gymnosperms and companion cells in flowering plants. The albuminous cells are rarely derived from the same precursor as the sieve cells, whereas companion cells are formed by divisions of sieve element precursors and sieve elements and companion cells are always ontogenetically related. The presence of companion cells is one of characteristic features of flowering plants (Eames and MacDaniels 1947; Esau 1977).

With the evolution of the sieve elements (see especially Esau 1977; and Esau, Cheadle, and Gifford 1953), the sieve areas become more and more differentiated. In the early stages of evolution, all the sieve areas of a particular sieve element are more or less similar, but afterwards the sieve areas with thicker connecting strands and more developed pores begin to separate. The sieve areas with more developed pores are usually localized on certain walls of the sieve elements, most often on their end walls. Portions of the wall bearing the more highly differentiated sieve areas with comparatively large pores are called sieve plates. The sieve plate consisting of many sieve areas (in scalariform or reticulate arrangement) is known as the compound sieve plate. If the sieve plate consists only of one sieve area, it is called the simple sieve plate.

The most primitive sieve tubes consist of long and narrow elements with strongly inclined, wedge shaped walls and with more or less similar sieve areas at the end and the lateral walls (Hemenway 1911; MacDaniels 1918; Esau, Cheadle, and Gifford 1953; Cheadle 1956; Zahur 1959; Easau 1977; Carlquist 1988). In the course of evolution, the end walls of the sieve-tube elements tend to change in position from very oblique to more or less transverse (perpendicular to the lateral walls). With the decrease in the inclination of the end walls, the number of sieve areas in the sieve plates diminishes. Simultaneously, a gradual reduction of the sieve areas on the lateral walls takes place. Thus, the most specialized sieve-tube elements are characterized by simple sieve plates with large pores on transverse

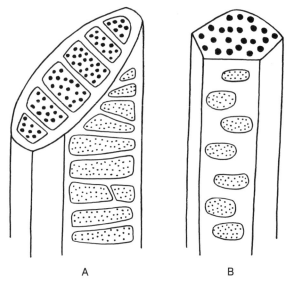

Figure 9. Stages in the evolution of a sieve plate from the compound to the simple (from K. Esau and V. I. Cheadle 1959).

A   B

end walls. The compound sieve plates pass into simple ones, more adapted to the function of conduction (figure 9; Cheadle and Whitford 1941; Esau and Cheadle 1959). This evolutionary process is similar to the transformation of the scalariform perforation into the simple one. Lastly, during evolutionary advance, a decrease in length and an increase in the diameter of the sieve-tube elements occured, which, however, did not lead here to those extremely barrel-shaped varieties with the diameter exceeding the height, as are found among some highly specialized vessel elements.

Thus, the evolution of the sieve tubes, as that of the vessels, was directed to the development of structures which are most suitable for the flow of liquid. In this context, some correlation is observed in the evolution of the sieve tubes and vessels and, as a rule, the degree of specialization of the sieve tubes more or less corresponds to the level of development of the vessels. But, there are also some nonconformities, as, for example, simple sieve plates and vessels with long scalariform perforation plates in *Cornus mas*. For the functional explanations of these nonconformities see Carlquist 1975:224.

1.6.3. Evolutionary Trends in Radial and Axial Parenchyma of Secondary Xylem and Phloem

The radial (transverse or horizontal) bands of the parenchyma cells called xylem (and phloem) rays are the oldest storage structures in the secondary conducting tissues of the higher plants. According to Kribs's (1935) terminology, individual rays may be either homocellular or heterocellular. Homocellular rays are composed of cells of one morphological type and consist of only procumbent or only square upright cells. Heterocellular rays consist of several tiers of procumbent cells flanked marginally by a series of upright cells which are elongated in the vertical direction. The entire ray system may consist either of homocellular or heterocellular rays only, or of various combinations of two types. When the ray tissue system consists of homocellular rays only, it is called homogenous; and when the rays are either all heterocellular or combinations of homocellular and heterocellular, the whole tissue system is called heterogenous (Kribs 1935; Esau 1977; Chalk, in Metcalfe and Chalk 1983). There are definite evolutionary trends in ray tissue system (Kribs 1935; Barghoorn 1941a, 1941b; Takhtajan 1959; Chalk, in Metcalfe and Chalk 1983; Carlquist 1988).

The most primitive type of ray tissue system is a heterogenous ray system which consists of two kinds of rays: one heterocellular—multiseriate composed of elongated or nearly isodiametric cells in the multiseriate part and upright cells in the uniseriate marginal parts which are longer than the multiseriate part; the other homocellular—uniseriate composed entirely of upright (vertically elongated) or of upright and square cells

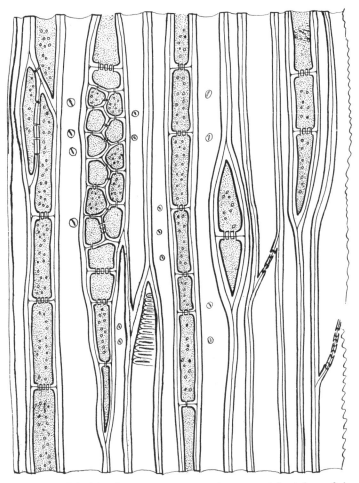

Figure 10. Primitive heterogeneous rays in tangential section of the wood of *Illicium parviflorum* (from A. Takhtajan 1948).

(figure 10). Such are, for example, the rays of the Winteraceae, *Eupomatia, Talauma, Illicium, Austrobaileya, Amborella,* Calycanthaceae, *Tetracentron, Platanus,* Actinidiaceae, *Davidia,* Aquifoliaceae, and others. This type of the heterogenous ray (also called Heterogenous I) dominates in stems with primitive vessels having scalariform perforations.

With increasing specialization of the heterogenous ray tissue system, multiseriate rays are made more fusiform in the tangential section, their uniseriate margins shortened, and the cells of uniseriate rays become shorter. According to Kribs (1935), this type of ray dominates in the stems having vessels with both the scalariform and simple perforations or with only simple perforations. It is found, for example, in a majority of Magnoliaceae, *Cercidiphyllum*, Buxaceae, Casuarinaceae, many Juglandaceae, and others. An intermediate type may be found, for example, in *Trochodendron*.

On subsequent evolution of the heterogenous rays, in some cases, the multiseriate rays were eliminated and, in others, the uniseriate rays disappeared. As a result, the uniseriate heterogenous and the multiseriate heterogenous rays arose. The uniseriate heterogenous type is found in the Myrothamnaceae, many Theaceae, and others. The multiseriate heterogenous type can be illustrated by the rays of woody Papaveraceae.

In the course of evolution, the heterogenous type of rays gave rise to the homogenous type. This characterizes woods with a simple perforation of vessels. The homogenous rays are found in three types—mixed homogenous, multiseriate homogenous, and uniseriate homogenous (figure 11).

All the diverse forms of distribution of the axial (longitudinal or vertical) parenchyma of the flowering plants are grouped into two main types—apotracheal and paratracheal (Chalk 1937, 1983). The apotracheal parenchyma is the most primitive type of axial parenchyma in the flowering plants. It is characterized by the fact that the arrangement of parenchyma cells does not depend on the arrangement of the vessels, although the parenchyma cells may often lie close to the vessels, particularly when the rows of parenchyma cells are numerous. The most primitive type of apotracheal parenchyma is that known as diffuse (Kribs 1937; Chalk 1983). This consists typically of individual cells (or strands) scattered throughout the fibrous ground tissue.

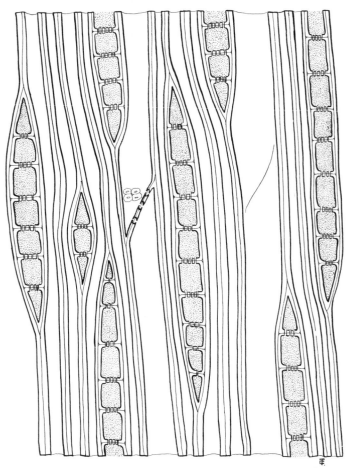

Figure 11. Specialized uniseriate homogeneous rays of *Populus euphratica* (from A. Takhtajan 1948).

About 90 percent of the woods with primitive vessels have diffuse parenchyma (Bailey 1957a; Chalk 1983). The diffuse type decidedly dominates in the stems with the scalariform perforation of vessels, not to speak of the vesselless stems. As woods became more specialized the proportion of axial parenchyma increased (Chalk 1983). This change led to the organization of the parenchyma in tangential lines instead of scattered cells. In such a case, the transverse section shows an aggregation of wood parenchyma as concentric (tangential) layers, mostly independent of the vessels. This is so-called metatracheal or banded parenchyma. All steps from the diffuse parenchyma to the metatracheal are found. The metatracheal parenchyma is met with in some archaic flowering plants like the genus *Talauma* of Magnoliaceae, but it is more characteristic of the specialized taxa.

The boundary or marginal parenchyma (Hess 1950) with single cells or a band at the end or at the beginning of the growth rings, originated from the diffuse parenchyma by reduction. Marginal parenchyma is already present in such archaic family as Magnoliaceae, but it is found more often in the specialized groups.

Paratracheal parenchyma (Sanio's term) is a more advanced type than the apotracheal. Here the parenchyma cells are more or less closely associated with vessels and other tracheary elements, which is a physiological progress (figure 12). Paratracheal parenchyma is divided into three main classes—vasicentric, consisting of a concentric sheath around vessels; aliform, with a winglike sheath projecting from the sides of a vessel; and confluent, in which the parenchyma around some of the vessels links up to form bands (Chalk 1983).

1.6.4. Origin and Evolution of Wood Fibers

The supporting system in the flowering plant wood consists of tracheids and/or fiber cells. In the most primitive woods, conduction and support are combined in tracheids. During evolution, the division of functions of conduction and strengthening occurred and the typical tracheids gave rise to more specialized in support fiber-tracheids, which in their turn gave rise to the libriform fibers (Jeffrey 1917; Bailey 1924, 1936, 1953; Chalk 1937, 1983; Esau 1977; Carlquist 1988). The evolution

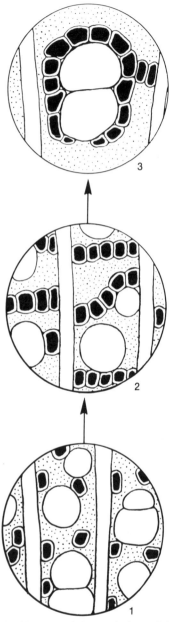

Figure 12. Schematic diagram of the evolution of the main types of wood parenchyma. 1, diffuse; 2, metatracheal; 3, vasicentric (from A. Takhtajan 1969).

from the tracheids to the libriform fibers was accompanied by an increase in the thickness of the wall (and a corresponding decrease in the diameter of the lumina), as well as by a reduction in the number and size of the pits.

An intermediate position between the tracheids and the typical libriform fibers is occupied by fiber-tracheids, which have fine bordered pits with well-defined contours. The fiber-tracheids are typically narrower than tracheids and their walls are usually thicker. They are usually associated with vessels having scalariform perforation and with heterogenous rays. The last stage of this evolution was the emergence of the typical libriform fibers characterized by thick walls and simple pits. These principally strengthening elements are longer than the fiber-tracheids.

Xylem fibers may be septate. Septation typically occurs in libriform fibers, but septation of fiber-tracheids is also known (Chalk 1983). Septate and nonseptate fibers most commonly occur together. Those that are septate are sometimes scattered among those that are nonseptate in a manner suggesting the distribution of diffuse parenchyma. More rarely, they are distributed in distinct bands resembling apotracheal parenchyma in cross-section, or in bands or mixed septate fibers and parenchyma (Chalk 1983). As opposed to the fibers, the septate fibers generally retain their protoplasts for a fairly long time, where reserves of starches as well as of oils and resins are deposited (Harrar 1946). There is an obvious resemblance between septate fibers and parenchyma not only in the occurrence of starch but in the ways in which they distributed (Chalk 1983). Septate fibers tend to be common throughout large natural groups, such as those which include the Meliaceae and Burseraceae, and "they are consequently a most useful index of affinity" (Chalk 1983:30).

68	*Vegetative Organs*

References

Arber A. 1950. The natural phylosophy of plant life. Cambridge.

Bailey I. W. 1924. The problems of identification of the woods of Cretaceous and later dicotyledons etc. Ann. Bot. 38:437–451.

Bailey I. W. 1936. The problem of differentiating and classifying tracheids, fiber tracheids, and libriform fibers. Trop. Woods 45:18–23.

Bailey I. W. 1944. The development of vessels in angiosperms and its significance in morphological research. Amer. J. Bot. 31:421–428.

Bailey I. W. 1949. Origin of the angiosperms; need for a broadened outlook. J. Arnold Arbor. 30:64–70.

Bailey I. W. 1953. Evolution of the tracheary tissues of land plants. Amer. J. Bot. 40(1):4–8.

Bailey I. W. 1956. Nodal anatomy in retrospect. J. Arnold Arbor. 37(3):269–287.

Bailey I. W. 1957a. The potentialities and limitations of wood anatomy in the study of phylogeny and classification of angiosperms. J. Arnold Arbor. 38:243–254.

Bailey I. W. 1957b. Additional notes on the vesselless dicotyledon *Amborella trichopoda* Baill. J. Arnold Arbor. 38:374–378.

Bailey I. W. and R. A. Howard 1941. The comparative morphology of the Icacinaceae. II. Vessels. IV. Rays of the secondary xylem. J. Arnold Arbor. 22:171–187, 556–568.

Baranova M. A. 1972. Systematic anatomy of leaf epidermis in the Magnoliaceae and some related families. Taxon 21:447–469.

Baranova M. A. 1975. Stomatographic investigation of the family Flagellariaceae. Bot. Zhurn. (Leningrad) 60:1690–1697. (In Russian.)

Baranova M. A. 1985. Classifications of the morphological types of stomata. Bot. Zhurn. 70(12):1585–1594. (In Russian.)

Baranova M. A. 1987a. On the stephanocytic type of the stomatal apparatus in Angiospermae. Bot. Zhurn. 72(1):59–62. (In Russian.)

Baranova M. A. 1987b. Historical development of the present classification of morphological types of stomates. Bot. Rev. 53(1):53–79.

Barghoorn E. S. 1940. The ontogenetic development and phylogenetic specialization of rays in the xylem of dicotyledons. I. Primitive ray structure. Amer. J. Bot. 27:918–928.

Barghoorn E. S. 1941a. The ontogenetic and phylogenetic specialization of rays in the xylem of dlcotyledons. II. Modification of the multi-seriate and uniseriate rays. Amer. J. Bot. 28:273–282.

Barghoorn E. S. 1941b. The ontogenetic and phylogenetic specialization of rays in the xylem of dicotyledons. III. The elimination of rays. Bull. Torrey Bot. Club 68:317–325.

Benzing D. H. 1967. Developmental patterns in stem primary xylem of woody Ranales. Amer. J. Bot. 54:805–820.

Bews J. W. 1927. Studies in the ecological evolution of the angiosperms. New Phytol. 26:1–21, 65–84, 129–148, 209–248, 273–294.

Bierhorst D. W. 1971. Morphology of vascular plants. New York.

Canright J. E. 1955. The comparative morphology and relationships of the Magnoliaceae. IV. Wood and nodal anatomy. J. Arnold Arbor. 36:119–140.

Carlquist, S. 1975. Ecological strategies of xylem evolution. Berkeley, Los Angeles, London.

Carlquist S. 1988. Comparative wood anatomy. Berlin.

Chalk L. 1937. The phylogenetic value of certain anatomical features of dicotyledonous woods. Ann. Bot. 1:409–428.

Chalk L. 1983. Wood structure. In C. R. Metcalfe and L. Chalk, eds., Anatomy of the dicotyledons. 2d ed. , pp. 2–51. Oxford.

Cheadle V. I. 1943a. The origin and certain trends of specialization of the vessels in the Monocotyledoneae. Amer. J. Bot. 30:11–17.

Cheadle V. I. 1943b. Vessel specialization in the late metaxylem of the various organs in the Monocotyledoneae. Amer. J. Bot. 30:484–490.

Cheadle V. I. 1944. Specialization of vessels within the xylem of each organ in the Monocotyledoneae. Amer. J. Bot. 31:81–92.

Cheadle V. I. 1953. Independent origin of vessels in the monocotyledons and dicotyledons. Phytomorphology 3(1):23–44.

Cheadle V. I. 1956. Research on xylem and phloem—progress in fifty years. Amer. J. Bot. 43(9):719–731.

Cheadle V. I. and N. B. Whitford 1941. Observations on the phloem in the Monocotyledoneae. I. The occurence and phylogenetic specialization in structure of the sieve tubes in the metaphloem. Amer. J. Bot. 28(8):623–627.

Corner E. J. H. 1949. The durian theory of the origin of the modern tree. Ann. Bot. 52:367–414.

Cotthem W. van 1970. A classification of stomatal types. Bot. J. Linn. Soc. (London). 63(3):235–246.

Cronquist A. 1968. The evolution and classification of flowering plants. London.

Cronquist A. 1988. The evolution and classification of flowering plants. 2nd ed. New York.

Davis P. H. and V. H. Heywood 1963. Principles of angiosperm taxonomy. Edinburgh and London.

De Bary A. 1877. Vergleichende Anatomie der Vegetationorgane der Phanerogamen und Farne. Leipzig.

Doyle J. A. and L. J. Hickey 1976. Pollen and leaves from the Mid-Cretaceous Potomac Group and their bearing on early angiosperm evolution. In C. B. Beck ed. Origin and early evolution of angiosperms, pp. 139–206. New York.

Eames A. 1911. On the origin of the herbaceous type in the angiosperms. Ann. Bot. 25:215–224.

Eames A. 1961. Morphology of the angiosperms. New York, Toronto, London.

Eames A. and L. H. MacDaniels 1947. An introduction to plant anatomy. 2d ed. New York and London.

Esau K. 1965. Plant anatomy. 2d ed. New York.

Esau K. 1977. Anatomy of seed plants. 2d ed. New York.

Esau K. and V. I. Cheadle 1959. Size of pores and their contents in sieve elements of dicotyledons. Proc. Nat. Acad. Sci. 45:156–162.

Esau K. , V. I. Cheadle, and E. M. Gifford 1953. Comparative structure and possible trends of specialization of the phloem. Amer. J. Bot. 40:9–19.

Ettingshausen C. von. 1861. Blattskelete der Dicotyledonen, Vienna.

Fahn A. 1974. Plant anatomy. 2d ed. Oxford.

Florin R. 1933. Studien über die Cycadales des Mesozoicums Erörterungen über die Spaltöffnungsapparate der Bennettitales. K. Svensk. Akad. Handl. 12:1–134.

Florin R. 1958. On Jurassic taxads and conifers from North-Western Europe and Eastern Greenland. Acta Horti Bergian 17(10):257–402.

Foster A. S. 1950. Morphology and venation of the leaf in Quiina acutangula Ducke. Amer. J. Bot. 37:159–171.

Foster A. S. 1961. The phylogenetic significance of dichotomous venation in angiosperms. Rec. Adv. Bot. 2:971–975.

Foster A. S. and E. M. Gifford, Jr. 1974. Comparative morphology of vascular plants. San Francisco.

Frost F. H. 1930. Specialization in secondary xylem of dicotyledons. II. Evolution of end walls of vessel segment. Bot. Gaz. 90:198–212.

Frost F. H. 1931. Specialization in secondary xylem of dicotyledons. III. Specialization of lateral walls of vessel segment. Bot. Gaz. 91:88–96.

Gamalei Yu. V. 1988a. The structural and functional evolution of leaf minor veins. Bot. Zhurn. 73:1513–1522. (In Russian with English summary.)

Gamalei Yu. V. 1988b. The taxonomical distribution of the leaf minor vein types. Bot. Zhurn. 73:1662–1672. (In Russian with English summary.)

Gamalei Yu. V. 1989. Structure and function of leaf minor veins in trees and herbs. Trees 3:96–110.

Golubev V. N. 1959. On morphogenesis of the woody plants and directions of morphological evolution from trees to herbs. Bull. Mosc. Obshch. Ispyt. Prirody 64(5): 49–60. (In Russian.)

Hallier H. 1901. Über Verwandtschafterhältnisse der Tubifloren und Ebenalen, etc. Abh. Naturwiss. Verein Hamburg 16(2):1–112.

Hallier H. 1902. Beiträge zur Morphogenie der Sporophylle und des Trophophylls in Beziehung zur Phylogenie der Kormophyten. Jahrb. Hamburgg wiss. Anst. XIX, 3. Beiheft, 1–110.

Hallier H. 1905. Provisional scheme of the natural (phylogenetic) system of flowering plants. New Phytol. 4(7):151–162.

Hallier H. 1912. L'origine et la système phylétique des Angiosperms exposée à l'aide de leur arbre généalogique. Archs. Néerl. Sci. sér. 2, 1:146–234.

Harrar E. S. 1946. Notes on starch grains in septate fibre-tracheids Trop. Woods 85:1–9.

Hemenway A. F. 1911. Studies on the phloem of the dicotyledons. I. Phloem of the Juglandaceae. Bot. Gaz. 51:131–135.

Hemenway A. F. 1913. Studies on the phloem of the dicotyledons. II. The evolution of the sieve-tube. Bot. Gaz. 55:236–243.

Hess R. W. 1950. Classification of the wood parenchyma in dicotyledons. Trop. Woods 96:1–20.

Hickey L. J. 1971. Evolutionary significance of leaf architectural features in the woody dicots. Amer. J. Bot. 58:469 (abstr.).

Hickey L. J. 1973. Classification of the architecture of dicotyledonous leaves. Amer. J. Bot. 60:17–33.

Hickey L. J. 1979. A revised classification of the architecture of dicotyledonous leaves. In C. R. Metcalfe and L. Chalk, eds., Anatomy of the dicotyledons, 2d ed. , Vol. 1, pp. 25–39, Oxford.

Hickey L. J. and J. A. Doyle. 1972. Fossil evidence on evolution of angiosperm leaf venation. Amer. J. Bot. 59:661 (abstr.).

Hickey L. J. and J. A. Wolfe. 1975. The bases of angiosperm phylogeny: vegetative morphology. Ann. Missouri Bot. Gard. 62:538–589.

Holttum R. E. 1955. Growth-habits of monocotyledons — variations on the theme. Phytomorphology 5(4):399–413.

Inamdar J. A., J. S. S. Mohann, and R. B. Subramanian. 1986. Stomatal classifications — a review. Feddes Repert. *97* (3–4):147–160.

Ishikawa M. 1918. Studies on the embryo sac and fertilization in *Oenothera*. Ann. Bot. 32:279–317.

Jeffrey C. 1917. Anatomy of woody plants. Chicago.

Kerner von Marilan A. 1887. Pflanzenleben. I. Leipzig.

Kribs D. A. 1935. Salient lines of structural specialization in wood rays of dicotyledons. Bot. Gaz. 96:547–557.

Kribs D. A. 1937. Salient lines of structural specialization in the wood parenchyma of dicotyledons. Bull. Torrey Bot. Club 64:177–186.

Lindley J. 1830. An introduction to the natural system of Botany. London.

MacDaniels L. H. 1918. The histology of the phloem in certain woody angiosperms. Amer. J. Bot. 5:347–378.

Mädler K. 1975. Ueber die Möglichketien einer plamässigen morphologischen Analyse der dikotylen Blätter. Courier Forschungsinstitut Senckenberg. 13:70–118.

Marsden M. P. F. and I. W. Bailey. 1955. A fourth type of nodal anatomy in dicotyledons, illustrated by *Clerodendron trichotomum* Thunb. J. Arnold Arbor. 36:1–50.

Melville R. 1976. The terminology of leaf architecture. Taxon 25:549–561.

Metcalfe C. R. and L. Chalk. 1950. Anatomy of the dicotyledons. 1, 2. Oxford.

Mouton J. A. 1970. Architecture de la nervation foliaire: Compt. Rend. 92 Congrès national de sociétés savantes (Strasbourg et Colmar, 1967)3:165–176.

Ozenda P. 1949. Recherches sur les Dicotylédones apocarpiques. Paris.

Pant D. D. 1965. On the ontogeny of stomata and other homologous structures. Plant Science Series, Allahabad 1:1–24.

Parkin J. 1953. The durian theory—a criticism. Phytomorphology 3:80–88.

Patel J. D. 1978. How should we interpret and distinguish subsidiary cells? Bot. J. Linn. Soc. (London) 77:65–72.

Patel J. D. 1979. A new morphological classification of stomatal complexes. Phytomorphology 29:218–229.

Rasmussen H. 1981. Terminology and classification of stomata and stomatal development—a critical survey. Bot. J. Linn. Soc. (London) 83(3):199–212.

Razdorsky V. F. 1955, Architecture of plants. Moscow. (In Russian.).

Serebryakov I. G. 1952. Morphology of vegetative organs of higher plants. Moscow. (In Russian.)

Serebryakov I. G. 1955. Main directions of the evolution of life forms in angiospermous plants. Bull. Mosc. Obshch. Ispyt. Prirody, Otd. biol. 60(3):71–91. (In Russian.)

Sinnot E. W. and I. W. Bailey. 1914. Investigations on the phylogeny of the angiosperms. 3. Nodal anatomy and the morphology of stipules. Amer. J. Bot. 1:441–453.

Sinnot E. W. and I. W. Bailey. 1915. Changes in the fruit type of angiosperms coincident with the development of the herbaceous habit. Report Bot. Soc. Amer. in Science 41:179.

Spicer R. A. 1986. Pectinal veins: a new concept in terminology for the description of dicotyledonous leaf patterns. Bot. J. Linn. Soc. 93:379–388.

Stearn W. T. 1966. Botanical latin. London.

Stebbins G. L. 1974. Flowering plants. Evolution above the species level. Cambridge, Mass.

Stebbins G. L. and G. S. Khush. 1961. Variation in the organization of the stomatal complex in the leaf epidermis of monocotyledons and its bearing on their phylogeny. Amer. J. Bot. 48:51–59.

Swamy B. G. L. 1949. Further contributions to the morphology of the Degeneriaceae. J. Arnold Arbor. 30:10–38.

Takhtajan A. 1943. Correlations of ontogenesis and phylogenesis in higher plants. Trudy Erevan State Univ. 22:71–176. (In Russian with English and Armenian summaries.)

Takhtajan A. 1948. Morphological evolution of the angiosperms. Moscow. (In Russian.)

Takhtajan A. 1959. Die Evolution der Angiospermen. Jena.

Takhtajan A. 1964. Foundations of the evolutionary morphology of angiosperms. Moscow and Leningrad. (In Russian.)

Takhtajan A. 1966. A system and phylogeny of the flowering plants. Moscow and Leningrad. (In Russian.)

Takhtajan A. 1969. Flowering plants. Origin and dispersal. Edinburgh.

Takhtajan A. 1980. Outline of the classification of flowering plants (Magnoliophyta). Bot. Rev. 46(3):225–359.

Thompson D'Arcy W. 1917, 1942. On growth and form. Cambridge.

Tomlinson P. B. 1974. Development of the stomatal complex as a taxonomic character in the monocotyledons. Taxon 23:109–128.

Troll W. 1939. Vergleichende Morphologie der höheren Pflanzen. Band 1, Teil 2. Berlin.

Wardlaw C. W. 1952. Phylogeny and morphogenesis. London.

Wardlaw C. W. 1965. Organization and evolution in plants. London.

Wilkinson H. P. 1979. The plant surface (mainly leaf). Part 1. Stomata. In C. R. Metcalfe and L. Chalk, eds., Anatomy of the Dicotyledons, Vol. 1, 2d ed., 97–117. Oxford.

Williams N. H. 1979. Subsidiary cells in Orchidaceae: their general distribution with special reference to development in Oncidieae. Bot. J. Linn. Soc. (London) 78:41–66.

Zahur M. S. 1959. Comparative study of the secondary phloem of 423 species of woody dicotyledons belonging to 85 families. Mem. Cornell Univ. Agr. Exp. Sta. 358.

Zhukovsky P. M. 1964. Botany. Moscow. (In Russian.)

2

Evolutionary Trends in Flowers and Inflorescences

2.1. *General Floral Structure*

The angiosperm flower emerged due to an aggregation of sporophylls in the distal parts of the shortened and determinate reproductive shoot, which provided a better protection to the young micro-and megasporangia and led to an improvement in the mechanism of pollination. The main distinctive feature of a flower compared to all known strobili of other seed plants is the presence of carpels, i.e., of more or less closed megasporophylls. The strobilus-like character of the flower is most evident in such archaic families as Magnoliaceae, Degeneriaceae, Winteraceae, Annonaceae, Nymphaeaceae, Ranunculaceae, Paeoniaceae, etc.

The most primitive flowers, like those of *Degeneria,* many Magnoliaceae, *Galbulimima, Eupomatia,* and Winteraceae, are of moderate size with a moderately elongated receptacle. Stebbins (1974) concluded that the original flowering plants had flowers of moderate size, which is in harmony with the hypothesis that they were small woody plants inhabiting pioneer habitats exposed to seasonal drought. It is also in harmony with my hypothesis of the neotenous origin of flowering plants, according to which they arose under environmental stress, probably as a result of adaptation to moderate seasonal drought on rocky slopes in an area with a monsoon climate (Takhtajan 1976). Large flowers, like those of some Magnoliaceae and Nymphaeaceae, of the Peruvian ranunculaceous *Laccopetalum giganteum,* and especially very large flowers (*Rafflesia*) are of secondary origin. Small and especially very small flowers are also derived. Their origin is usually correlated either with the specialization of inflorescences or with the reduction of the whole plant.

2.2. From Spiral to Cyclic Flowers

The most primitive flowers have a more or less indefinite and variable number (but not necessarily a large number) of separate parts arranged spirally upon a moderately long floral axis. The progressive shortening of the floral axis brings free floral parts closer together and gives rise to the gradual transition from spiral to cyclic arrangement and to the fixation of the number of parts. At its earlier evolutionary stages, this progressive shortening is reversible, and in some relatively archaic taxa, such as Magnoliaceae (especially *Magnolia pterocarpa*), *Schisandra,* or *Myosurus,* the elongated receptacle is of secondary origin. The elongated receptacle is a specialization related to dispersal mechanisms (Carlquist 1969:335).

The cyclic arrangement is more parsimonious and "had the adaptive value both of placing the stamens and carpels near to each other, thus increasing the efficiency of insect pollination, and speeding up floral development, in response to the demands of a seasonally unfavorable climate" (Stebbins 1974:283). Cyclic flowers are already found in such relatively archaic plants as some Annonaceae, some Lauraceae, *Austrobaileya, Lactoris,* Aristolochiaceae, some Ranunculaceae (*Aquilegia*), etc. They are dominant among the higher orders of the Magnoliophyta, including the whole class of Liliopsida.

The transition from the spiral anthotaxis to the cyclic arrangement begins either from the perianth or from the carpels, but in some cases it begins simultaneously from both ends. Most often, as in the majority of Magnoliaceae and Annonaceae, only the stamens and carpels are arranged spirally, whereas the perianth is more or less cyclic (mostly in trimerous whorls, as, e.g., in *Magnolia denudata*—see Erbar and Leins 1981). The same is the case with numerous spirocyclic representatives of the Ranunculaceae (e.g., in *Eranthis* and *Ranunculus*) and *Hydrastis*. But in *Amborella,* the arrangement of both the stamens and carpels is cyclic; and in *Illicium* and *Sargentodoxa,* only the carpels are arranged in cyclic manner. Lastly, in some Annonaceae (*Isolona* and *Monodora*) and in *Glaucidium,* only the arrangement of stamens is spiral.

But the transformation of the spiral arrangement into the cyclic is not always undirectional and in some cases, especially in relatively archaic dicotyledons, anthotaxis is more or less reversible. According to Endress (1987), at the lower evolutionary level of the dicotyledons, the change between spiral and cyclic (mainly trimerous) arrangement is easy and a switch to and back again must have occurred repeatedly, at least in some families, such as Ranunculaceae or Menispermaceae. In the apetalous male flowers of some species, there may be also a switch from the cyclic to spiral or unordered patterns (Endress 1987), as for instance in *Humulus lupulus* (Leins and Orth 1979) or in *Quercus rubra* (Sattler 1973). But the general direction of evolution is from a spiral to a cyclic arrangement.

2.3. Oligomerization of the Homologous Flower Parts

The flowers of some Magnoliidae and Ranunculidae are still more or less polymerous and consist of an indefinite number of parts. But going up the phylogenetic ladder, we clearly observe an increasing oligomerization of the flower and a fixation of the number of homologous parts connected with it. All of the most advanced flowering plants, such as Rubiales, Solanales, Lamiales, Campanulales, Asterales, etc., are characterized by a considerable oligomerization of the flower and a very strict fixation of both the number of its whorls and the number of parts of each whorl. Only relatively rarely this general trend of an increasing of oligomerization is replaced by polymerization, as we find it, for example, in the Aizoaceae and Cactaceae and—most probably also—in some Magnoliaceae and Nymphaeaceae.

Flowers with a small and fixed number of parts are more integrated than archaic, less oligomerized flowers. The structural integration is accompanied by an integration of functions so that the wholeness of the flower increases with evolution. Consequently, the general rate of development of the flower and the fruit increases greatly.

The evolutionary significance of the principle of oligomerization was expressed in botany by Nägeli (1884) and then stressed by Merezhkovsky (1910:165) who formulated "the law of integration of the homologues." According to Merezhkovsky, a large number of the parts of the primitive types of the flowers "corresponds to the general principle of evolution and particularly to the law of integration of homologues, in conformity with which the number of homologous parts or homologues decreases in the course of evolution."

2.4. Origin and Evolution of the Perianth

Among extinct gymnospermous plants only the strobili of Cacadeoideales (Bennettitales) had envelopes or perigonii, and among the extant phyla they developed only in *Ephedra, Welwitschia,* and *Gnetum,* the probable descendants of Cycadeoideales. The envelope of the strobili of both Cycadeoidales and Gnetophyta is "simple." It is undoubtedly of foliar origin. But in an overwhelming majority of Magnoliophyta, including many of their archaic members, the perianth (flower perigonium) is "double," i.e., consists of a calyx and a corolla. Thus, the problem of the origin of the angiosperm perianth is much more complicated than that of the envelope in the gymnospermous plants.

The most primitive angiosperm perianth is composed of morphologically equivalent tepals of bracteal origin, as in many Magnoliidae. These bractlike perianth parts emerged from the reduced and modified uppermost protective leaves (hypsophylls). This primitive perianth is more or less conspicuously colored and is usually differentiated into proximal and distal parts, the distal tepals larger and more petaloid. The family Magnoliaceae itself shows the gradual differentiation of the perianth into calyx and corolla. In *Michelia mannii* and in some species of *Magnolia* (especially in *M. quinquepeta*), the outer tepals are much reduced in size and sepal-like. In many other Magnoliaceae, all stages in the differentiation of the perianth are present. A perianth of completely bracteal origin is characteristic for Degeneriaceae, Magnoliaceae, Annonaceae, Canellaceae, Winteraceae, Illiciaceae, Schisandraceae, Austrobaileyaceae, Calycanthaceae, and other relatively archaic families (Smith 1928; Eames 1961; Hiepko 1965b). Petals in these families are "bracteopetals" (Kozo-Poljanski 1922).

But not all petals are "bracteopetals." It is almost generally agreed that they are of dual origin (Arber and Parkin 1907; Kozo-Poljanski 1922; Sprague 1925; Eames 1961; Hiepko 1965b; Cronquist 1968, 1988; Bierhorst 1971; Stebbins 1974; and others). In the overwhelming majority of Magnoliopsida and within the whole class Liliopsida, petals originated from sterilized and modified stamens, they are "andropetals." This hypothesis of staminal origin of petals in the majority of flowering plants received widest recognition, especially because of the works of Čelakovsky (1896), Worsdell (1903); Troll (1927, 1928), and Eames (1931).

The staminate nature of the petals is very clear in the members of the Menispermaceae, Ranunculaceae, Papaveraceae, Aizoaceae, Cactaceae, Caryophyllaceae, Capparaceae, Rosaceae, Tiliaceae, and many other families. In many cases, there are convincing anatomical data confirming the staminal origin of the petals. Thus, the staminal origin of the petals of the Ranunculaceae is proved by a study of the vascular system (Smith 1926, 1928) and comparative morphology and ontogeny (Hiepko 1965b), as well as by study of spatial relations between stamens and petals (petals and stamens are arranged on the same parastichies—Tamura 1965*) and by teratological data (including numerous cases of the substitution of stamens by petals) (Worsdell 1903).

But what are the evolutionary relationships between the bracteopetalous and andropetalous perianths? Bracteopetals occur in more archaic taxa and evidently appeared earlier. They are also connected with generally more primitive pollination mechanisms and with less specialized pollinators. Andropetals, on the contrary, are usually connected with more advanced

---

*On the contrary in *Paeonia*, which has been formerly included in the Ranunculaceae, "sepals and petals are arranged on the same phyllotactic spiral, and the transitional forms between them appear" (Tamura 1963:117). Morphologically, the petals of *Paeonia* are closely related to the sepals (Hiepko 1965a) and they are true bracteopetals.

types of pollination. But how could the andropetals arise if there already existed bracteopetalous flowers? It is very unlikely that in the course of evolution the bracteopetalous perianth was replaced by the andropetalous one. It could happen only in very special cases connected with very special kinds of pollination mechanism. It is much more probable that andropetals arose at a very early evolutionary stage of flowering plants and in flowers with a still very primitive simple perianth composed of bractlike sepals. The staminal petals turned out to be more prospective than the bracteopetals. The andropetals had higher evolutionary plasticity and were more liable to various kinds of differential growth. They are more neotenic than any other floral organ and, therefore, more plastic and adaptable. According to Goebel (1933), petals are stamens with arrested development, which explains their relatively late appearance in ontogeny.

Evolutionary specialization of the corolla, beginning with free petals, has given rise to a sympetalous corolla. A tendency to sympetaly is observed already in certain orders of the choripetalous magnoliophytes, such as Caryophyllales, Saxifragales, Rosales, and Geraniales, but the sympetalous corolla is characteristic mainly of the most advanced orders of the subclasses Lamiidae and Asteridae. With the specialization of the perianth, the bases of the synsepalous calyces and the sympetalous corollas fuse at a greater or lesser height and form the floral tube.

The petals in many flowering plants are more or less reduced or even completely absent. This is generally connected with the adaptation to wind pollination or self-pollination or, more rarely, with the transfer of their functions to the stamens or other organs.

There are also some definite trends in the aestivation (praefloration) of parts of the perianth, which for calyx and corolla can be the same or different within a flower. The basic and most primitive type of aestivation is the imbricate aestivation (Arber and Parkin 1907; Takhtajan 1948). There are different forms of imbricate aestivation. The most primitive is simply imbricate when the perianth parts overlap each other parallell at the margins, without any involution. It is characteristic for the spiral perianth and for the more primitive types of cyclic floral envelopes, e.g., for the perianth of the Magnoliaceae, Winteraceae, Nymphaeaceae, Theaceae or Liliaceae. A very common type of imbricate aestivation is the quincuncial, when the pieces (sepals or petals) are five in number, of which two are exterior, two interior, and the fifth covers the interior with one margin and has its other margin covered by the exterior, as in calyx of some Caryophyllaceae. Another form of imbricate aestivation is the contorted or convolute, when the right (or left) margin of each piece (sepal, petal, or their lobe) is overlapping the succeeding piece, as in corolla of Polemoniaceae or of *Convolvulus*.

Amongst the nonimbricate forms of aestivation, the commonest is the valvate, when the margins of sepals, petals, or their lobes meet without overlapping. The valvate aestivation occurs only in true cycles (whorls) and is common both in archaic families such as Winteraceae, Annonaceae, Myristicaceae, Lardizabalaceae, and in advanced families such as Loranthaceae, Viscaceae, Grubbiaceae, Rhizophoraceae, Onagraceae, Elaeagnaceae, Proteaceae, Campanulaceae, and Asteraceae.

There are also some other special forms of aestivation, such as induplicate (having the margins bent abruptly inwards, as in the flowers of some species of *Clematis*) or plicate (when the perianth is folded several times, as in *Nicotiana*).

2.5. Evolutionary Trends in Stamens and Androecium

2.5.1. Stamens

Comparative studies of the stamens of flowering plants lead to the conclusion that the most primitive type of stamen is a broad, usually more or less laminar, three-veined organ not differentiated into filament and connective and continued above two pairs of slender vertically elongate microsporangia situated between the lateral veins and the midvein (Hallier 1903a, 1903b, 1912; Arber and Parkin 1907; Parkin 1923, 1951; Bailey and Smith 1942; Bailey and Nast 1943; Bailey and Swamy 1948; Ozenda 1949, 1952; Swamy and Bailey 1950; Canright 1952; Moseley 1958; Takhtajan 1959; Eames 1961; Cronquist 1968, 1988; Bierhorst 1971; Foster and Gifford 1974; Endress and Hufford 1989). In a number of various magnoliids, especially in *Degeneria, Galbulimima, Magnolia maingayi,* and *Manglietia forrestii,* as well as in the species of *Elmerillia,* certain species of *Talauma,* in *Austrobaileya, Nymphaea, Victoria,* and *Nuphar,* and in certain other archaic dicotyledons the stamens are still of a relatively primitive type.

Already, Hallier (1903a, 1903b, 1912) and Arber and Parkin (1907) had suggested that the continuation of the connective beyond the anther as a sterile tip is a primitive feature. Later, Parkin (1923, and especially 1951) stressed this further. But the protrusion of the connective is intimately connected with pollination and in certain instances this protrusion "may have been elaborated to serve biological purposes, particularly pollination" (Parkin 1951 : 8). Therefore in some cases, as for example in the Magnoliaceae (Howard 1948), some monocotyledons, and even the Asteraceae, there is evident secondarily elongated protrusion of the connective, which is "an organ to serve some biological purpose" (Parkin 1951:6), most probably in connection with pollination. This is evident in the Malayan magnoliaceous genus *Aromadendron,* where the connective produced into a very long setaceous appendage subequaling or longer than the anther (Dandy, in Hutchinson 1964). Still more evident is the secondary character of protrusion of the connective in some members of Trilliaceae, especially in *Daiswa thibetica,* where the free portion of connective reaches 12 mm and sometimes even

17.5 mm (Takhtajan 1983). In the last case, it is clearly a derived character.

In primitive *Degeneria*-type stamens, the microsporangia are usually more or less embedded in the tissue of the stamen. In *Degeneria* and *Galbulimima,* the microsporangia are deeply sunk, as they are in the Magnoliaceae (except *Liriodendron*) and *Victoria amazoniaca.* Many authors, among them Ozenda (1952), Canright (1952), Moseley (1958), and Eames (1961) consider the immersion of the microsporangia as a primitive feature. But in the opinion of Willemstein, the immersion of the microsporangia "is not plesiomorphous, because it is either an adaptation to avoid too much self-pollination in protandrous, introrse conditions, or it is an adaptation to avoid too much pollen-feeding by the insect visitors. The conditions in, e.g., *Degeneria* (fleshy stamens) is paralleled in the Bennettitales in *Williamsoniella coronata* (fleshy microsporophylls) in clearly cantharophilous circumstances" (1987:269). The immersion of anthers is most probably an old adaptation, which we find only in archaic magnoliopsids. But, of course, it does not mean that the immersion is a primitive character per se. The immersion was more probably an ancient specialization for protection of anthers in primitive cantharophilous flowers.

In archaic families with primitive stamens, the arrangement of microsporangia is usually more or less laminar (figure 13). But while in *Degeneria, Galbulimima, Liriodendron,* Annonaceae, Canellaceae, Myristicaceae, some Winteraceae, Calycanthaceae, *Lactoris,* Lardizabalaceae, etc., the microsporangia occupy the abaxial (dorsal) surface, in Magnoliaceae (except *Liriodendron*), *Austrobaileya,* and Nymphaeaceae, they occupy the adaxial (ventral) surface. The presence of two diametrically opposite types of arrangement of microsporangia in taxa so close to each other, as for example, Magnolioideae and Liriodendroideae of Magnoliaceae and Himantandraceae, seems morphologically puzzling.

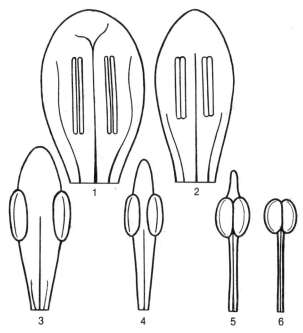

Figure 13. Main trend of evolution in stamens from the most primitive broad laminar type (from A. Takhtajan 1954).

From the fact that abaxial arrangement of sporangia characterizes the majority of ferns and the abaxial arrangement of microsporangia is characteristic for cycads, Hallier (1903a, 1912) considered the abaxial position of the anthers more primitive than the adaxial, which seems plausible (see also Chupov 1987). If we take into consideration the very high evolutionary plasticity of the stamens, and the flexibility of the orientation of microsporangia toward either the ventral or dorsal side of the stamens, we can easily assume an origin of the adaxial position from the abaxial one by way of merely a small shift in the site of meristematic activity of primordia, i.e., through archallaxis. Even a simple ontogenetic change could govern the position (Cronquist 1988:207).

The general evolutionary trends in the stamens were the

narrowing of the lamina, the differentiation of the lamina into filament and connective, the reduction of the number of veins to one, and the suppression of the protruded part of the connective.

In the primitive stamens, the microsporangia lie in two pairs that are separated by a more or less considerable portion of the sterile tissue. But with the evolution of the stamen, the lamina becomes narrower and narrower, its tissue is reduced, and—at maturity—the two microsporangia of each pair become confluent owing to the breaking down of the partition between them. When the lamina is reduced still more and the connective gradually disappears, the two bisporangial anther halves merge into one tetralocular anther which, in most cases, occupies the apex of the stamen. More than four locules or a lesser number of them is rarely found. A decrease in the number of locules occurs either due to the branching of the stamens or to the fact that certain locules do not develop. But the increase in the number of locules takes place owing to their separation by the layers of the sterile tissue.

In the beginning, the differentiated anther is basifixed, that is, attached proximally to the filament. A more advanced position of the anther is present when it is attached dorsally to the filament and thus the proximal part of the anther becomes detached from the filament. If a considerable part of the anther (not less than a third) is free from the filament, the anther is versatile and is capable of a swinging motion when touched by an insect or by wind.

2.5.2. Androecium

In some more archaic families like the Magnoliaceae, Eupomatiaceae, the majority of Annonaceae, the Illiciaceae, Austrobaileyaceae, Trimeniaceae, Nymphaeaceae, Nelumbonaceae, the majority of Ranunculaceae, the Hydrastidaceae, and the

Glaucidiaceae, the stamens are still spirally arranged and their number is more or less indefinite. Their androecium is primitively polystemonous ("polyandrous").

In most magnoliophytes, the stamens are arranged in whorls and thus the androecium is cyclic. The stamens in cyclic androecium form one or two whorls, very rarely several whorls. There are two basic types of the cyclic androecium—diplostemonous and obdiplostemonous. In the diplostemonous androecium, stamens of the outer whorl alternate with the petals (and opposite the sepals), while those of the inner whorl are opposite the petals. The diplostemonous androecium is the most common and more primitive type, which evolved directly from the polymerous spiral androecium by reduction and cyclisation. In the less common obdiplostemonous androecium, stamens of the outer whorls are opposite the petals, while those of the inner whorl alternate with the petals. Obdiplostemonous androecia are characteristic of some Caryophyllaceae, Clethraceae, Ericaceae, some Tamaricaceae, some Cunoniaceae, Saxifragaceae, Grossulariaceae, some Rutaceae, Zygophyllaceae, Peganaceae, Oxalidaceae, Geraniaceae, and some other taxa. Various interpretations of obdiplostemony were proposed (see especially Čelakovsky 1875, 1878, 1894; Eichler 1878; Stroebl 1925; Goebel 1928; Corner 1946; Eames 1961; Eckert 1966; Ronse = Decraene and Smets 1987). According to Stroebl (1925), Leins (1964a), Gelius (1967), Eckert (1966), Mayr (1969), and Ronse Decraene and Smets (1987), obdiplostemony originated as a result of differential growth during the development of the flower which brings about displacements (changes in direction of staminal growth, strong growth of the calyx, etc.). Therefore, the violation of Hofmeister's "Alternanz-Regel" is, in reality, only apparent.

In the majority of flowering plants, the cyclic androecium consists of only one whorl of stamens (haplostemonous and obhaplostemonous androecia). In the haplostemonous androecium, stamens alternate with the petals. The haplostemonous androecium is characteristic for many families including the whole subclasses Lamiidae and Asteridae. In the obhaplostemonous androecium, stamens are opposite the petals. Examples of obhaplostemony are Balanophoraceae, Plumbaginaceae, Santalales, Rhamnaceae, Vitaceae, and Haemodoraceae. Both haplostemonous and obhaplostemonous androecia originated from the diplostemonous androecium as a result of reduction of one of two whorls. In some cases, as for example in *Lysimachia vulgaris* and *Samolus,* there still occur small staminodia, remnants of the lost whorl.

Oligomerization of the androecium is the main, but not the only, trend in its evolution. In many taxa of flowering plants, we observe the polymerization of the cyclic androecium, providing greater pollen production, especially in large flowers. As a result, we have many types of secondary polymerous androecia which are completely different from the primitive polymerous androecia of archaic taxa. The multiplication of stamens is of two kinds. In some taxa, as for example in *Rosa,* the number of the stamen primordia is multiplied. The stamen number increases by way of the acropetal insertion of new whorls of primordia, coinciding with an expansion of the receptacle (Kania 1973). But usually, secondary polymery depends on the splitting of stamen primordia or of very young stamens rather than a real increase in the number of stamens. In the case of splitting, the androecium is fasciculate (each fascicle corresponding to an original stamen). Polymerous androecia evolved independently and heterochronously from a diplo- or haplostemonous state.

During evolution changed not only the number and arrangement of stamens but also the mode of their sequence of ontogenetic development (Payer 1857, Corner 1946). The initial and most widespread type of development is the centripetal (acropetal), when the development of androecium follows the development of the perianth in the normal sequence, spiral or cyclic. The first to develop in this case are the outermost (lowermost) stamens and then, successively, the inner ones. This type is characteristic for all spiral androecia (like those of Magnoliaceae, Annonaceae, Nymphaeaceae, Nelumbonaceae, Ranunculaceae), for cyclic oligomerous androecia, and for some cyclic polymerous androecia, such as those of the Papaveraceae, Rosaceae, Fabaceae-Mimosoideae, or Myrtaceae.

In the centrifugal androecium, there is a break between the order of development of the perianth and androecium caused by the intercalation of new stamens. The centrifugal development arose from the centripetal (Corner 1946; Ronse Decraene and Smets 1987). It is characteristic of the Glaucidiaceae, Paeoniaceae, probably some Phytolaccaceae with numerous stamens, Aizoaceae, Cactaceae, Dilleniaceae, Actinidiaceae, Theaceae, Clusiaceae, Lecythidaceae, many Violales, some Capparaceae, Bixaceae, Colchlospermaceae, Cistaceae, Tiliaceae, Bombacaceae, Malvaceae, the genus *Lagerstroemia* (Lythraceae), Punicaceae, Loasaceae, Limnocharitaceae, and some other taxa. In some families such as Ochnaceae, Begoniaceae, Lythaceae, and Loasaceae, there are both types of stamen development. Therefore, the distinction between centrifugal and centripetal types of development is by no means clear-cut and there are some transitional forms (Sattler 1972; Philipson 1975; Sattler and Pauzé 1978; Ronse Decraene and Smets 1987). According to Leins (1964b, 1975), the difference between centripetal and centrifugal development depends on the shape of the receptacle: a concave receptacle would give rise to a centripetal development, while on a convex receptacle only a centrifugal development would be possible. But this is not a general rule (Hiepko 1964, Mayr 1969, Ronse Decraene and Smets 1987).

The staminal filaments often fuse with each other, as well with the perianth parts and the carpels. In some cases, stamens coalesce so closely that it is difficult to separate them from each other, as for example in *Cyclanthera* (Cucurbitaceae). The filaments often coalesce into the staminal tubes surrounding the gynoecium as in the Malvaceae.

Another trend of evolutionary specialization of androecium is the transformation of some of fertile stamens into sterile staminodia. This transformation may occur either in inner or outer members. Transformation of the inner stamens seems to be the more primitive condition (Eames 1961). But in some cantharophilous archaic families such as Degeneriaceae, Himantandraceae, Eupomatiaceae, some Monimiaceae, Calycanthaceae, and Nymphaeaceae, staminodia are present both above and below the stamens. "If the upper position is the earlier one in the evolution of the flower, as suggested by these primitive taxa, the first corolla was above the stamens, a position which accompanies pollination by beetles," states Eames (1961:104).

2.6. Evolutionary Trends in Carpels and Gynoecium

2.6.1. Initial Stages of Evolution of Carpels

The carpels (megasporophylls) of the magnoliophytes are still more modified and specialized than their stamens. With evolution, they undergo considerably larger changes (reduction, concrescence, differentiation, etc.) than other parts of the flower. Therefore, the morphological interpretation of the carpels involves great difficulties and continues to cause a great divergence of opinion.

The angiosperm carpels originate from the leaflike open magasporophylls of the hypothetical gymnospermous ancestors, i.e., they have a phyllome character. The phyllome character of the carpels is most clearly evident in the primitive flowers of certain archaic taxa, especially in *Degeneria* and *Tasmannia*. Even in *Caltha,* a young carpel is very similar to a young leaf. The more primitive the flower, the more evident the leafy nature of the individual carpel. In contrast, as the flower evolved, the carpels gradually lost their original phyllome character and acquired a large number of specific features.

Primitive carpels very much resemble phyllome organs in their anatomical characters. As was shown by the investigations of Troll (1939) and his students, the phyllome character of the carpels is fully confirmed by the study of their histogenesis which does not differ in principle from the histogenesis of phyllome organs. The vascular structure of the carpels is also a very convincing proof of their phyllome character (see especially Eames 1931, 1961). Lastly, the leafy nature of the carpels is also confirmed by the atavistic anomalies, repeatedly described in literature.

Carpels (as well as stamens) originated not from the vegetative plotosynthesizing leaves (trophophylls), but from the spore-bearing phyllomes (sporophylls). The resemblance to the vegetative leaves is explained by the fact that both the trophophylls and the sporophylls have a common origin from the undifferentiated "trophosporophylls" of the type of many ferns and seed ferns (Potonié 1912:227) as well as by a certain parallelism in the evolution of the leaves and carpels (for example, in the evolution of their nodal anatomy and venation). But both carpels and the stamens emerged not directly from undifferentiated trophosporophylls, but from sporophylls. They are modified megasporophylls and microsporophylls.

The megasporophylls of the hypothetical ancestors of magnoliophytes could not have been so reduced as in the Cycadeoideales. Most probably, they were relatively large, with a fairly wide and possibly lobed or even branched blade and with a branched vascular system. The ovules in such ancestral megasporophylls had probably a marginal or submarginal arrangement. But how is it possible to conceive the origin of the closed and specialized carpels of the magnoliophytes from the primitive open megasporophylls of the gymnospermous ancestors?

The carpels of the most archaic magnoliophytes, especially those of *Degeneria* and *Tasmannia*, have a clearly expressed appearance of juvenilized, infantile structures. They are conduplicate during the early stages of their development and thus closely resemble young leaves folded adaxially along the midrib (Bailey and Nast 1943; Bailey and Swamy 1951). Like primitive stamens and many bracts and cataphylls, they are characterized by palmate venation. It was therefore not difficult to conclude that the evolutionary transformation of the ancestral open megasporophyll into the carpel by folding and closure along its midrib could have occured in a juvenile stage of ontogeny (Takhtajan 1948, 1959, 1969, 1976). Certainly, such a neotenic method of origin of carpels may be assumed only if the megasporophylls of the gymnospermous ancestor were also conduplicate in the vernation and not circinate as in seed ferns (including *Caytonia*). In this connection, it is important that conduplicate vernation is well known both in Cycadales and Cycadeoideales. Thus, it could have existed also in the ancestors of flowering plants.

Among the living magnoliophytes the most primitive type of carpel is found in *Tasmannia piperita* and allied species and in *Degeneria* (Bailey and Smith 1942; Bailey and Nast 1943; Swamy 1949; Bailey and Swamy 1951; Eames 1961). The carpels of these archaic flowering plants are stipitate, resembling a petiolate leaf (which is considered as a primitive feature—see Eames 1961:426) and have a conduplicately folded lamina. Three independent veins pass along the lamina, and the median (dorsal) of them forms extensive branches, whereas the two lateral (ventral) veins have shorter branches that extend both inward toward the median system and outward toward the free margins of the carpel. The numerous anatropous ovules are attached between the median and lateral veins. These primitive carpels are not differentiated into closed ovary, stylodium, and a localized stigma.

The primitiveness of such carpels is underlined by the fact

that their free margins (more or less conspicuously flaring in *Degeneria*), except the basal portion, are not completely closed at pollination and even later are joined only by the closely interlocking papillose epidermal hairs. These hairs are widely distributed both over the inner surface and the free carpellary margins and sometimes even extend outward onto more exposed outer surface. They represent an unspecialized stigmatic surface. At anthesis, pollen is retained by the external stigmatic hairs, and the pollen tubes penetrate among the more internal hairs into the cavity of the carpel. Following pollination and fertilization the adjacent margins of the carpel become sealed.

Incomplete concrescence of the carpellary margins is observed not only in the Magnoliidae, but also in certain other more advanced taxa. Thus in *Paeonia*, the margins of the carpel do not yet coalesce along their entire length, though the close contact has already been achieved between the border zones. In this case, the contacting epidermal layers remain still histologically well expressed for a long time. Such a state is observed also in many other taxa, for example in *Trochodendron, Caltha, Trollius, Delphinium, Hydrastis, Physocarpus, Spiraea, Sedum, Butomus, Tofieldia, Chamaerops*, etc. In the carpels of still more advanced genera, the double epidermal layer between the fused borders is already usually indistinct. At the first stages of concrescence, the lateral veins of the carpel converge but still remain independent. With further evolution, they often merge into a common sutural bundle (see Smith 1928; Chute 1930; Eames 1931, 1961).

With increasing specialization of the carpel and concrescence of its margins, the wide areas between the ovules and the borders, i.e., conduplicate ventral parts, start reducing gradually. Various stages of this process may be traced in the family Winteraceae. As a result, carpels appear whose margins seems to be curved inside (involute) and the placentation appears to be marginal. But, as Bailey and Swamy (1951) indicated, the

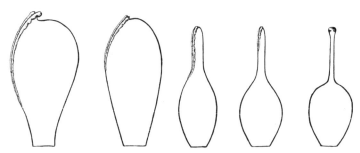

Figure 14. Main evolutionary trend in carpels from the most primitive conduplicate type (from A. Takhtajan 1954).

appearance of infolding (involution) and marginal placentation is due to inwardly projecting stalks of the ovules or to similarly oriented placental ridges. In other words, the placentation in such carpels is modified laminar, rather than truly marginal. This is the most common type of carpel. With the transition of the laminar placentation into the submarginal, the number of ovules inside the carpel decreases, and this tendency towards diminution of their number subsequently continues. A general decrease in the number of ovules inside the carpel may be observed, for example, within the Ranunculaceae.

With the emergence of the carpel, even such a primitive one as in *Tasmannia piperita* and *Degeneria,* the direct access of the pollen grains to the ovules is rendered difficult. Therefore, simultaneous with the emergence of the carpel is also formed a special stigmatic surface, which is capable of receiving the pollen grains and helping in the development of the pollen tube. Presence of the stigmatic surface (at first very primitive and then localized and specialized) is one of the most characteristic features of the angiosperm megasporophylls.

Initially the stigmatic surface stretches in a fairly wide band along the whole border zone of the carpel ("decurrent stigma") but then it is gradually localized and converted into a specialized stigma (figure 14). Thus, the stigma developed from carpellary margins converging during evolution and then coalescing. As

Kozo-Poljanski (1922:121) first pointed out in his commentary on Hallier's codex of the primitive characters, "the stigma developed from the sutures." Robert Brown (1840) had already drawn attention to this dual nature of the longitudinal stigmatic crest, basing his view very aptly on the structure of the carpel of the genus *Tasmannia*. Hallier (1912) was the first who recognized the primitiveness of decurrent stigma.

In the genus *Degeneria*, the stigmatic surface stretches over the length of the carpel. The stigmatic areas are not localized externally upon the flaring carpellary margins, but extend inward along the adaxial surface of the carpel into close proximity to the placentas. This is still a very primitive stigmatic surface, which is quite far from the localized and specialized stigma of the most advanced groups. In the species of the genus *Tasmannia*, the stigmatic surface also extends from the region of the stipe along the conduplicate adaxial parts of the carpel and slightly overtops its apex, but in the species of the genus *Drimys sensu stricto*, it is limited by the subapical portion of the carpel —i.e., more restricted and localized (Bailey and Nast 1943). There are quite primitive stigmas also in certain other representatives of the Winteraceae, as well as in *Schisandra*, *Cercidiphyllum*, *Euptelea*, and *Platanus*. But more often, and even in the archaic taxa, the stigmas are strictly localized in the apical part of the carpel (figure 14).

As the stigmatic area is localized in the upper part of the carpel, the latter is usually elongated into a thin sterile style-like excrescence, raising the stigma over the fertile portion of the carpel and serving as a path for growing pollen tube and a source of nutrients (figure 14). Hanf (1935) aptly proposed to call it stylodium. Usually the stylodium is called "style" which is somewhat incorrect. Only when styllodia fuse together is a true style formed. "There may be all gradations between completely separate stylodia and fully fused ones. As an alternative, stylar structures could be spoken of as simple or compound styles.

But the introduction of stylode (stylodium) for the former and style only for the latter might be preferable" (Parkin 1955:51). During the earlier stages of its evolution, the stylodium is still conduplicate with a distinct ventral groove and a decurrent stigma, consisting of two relatively wide stigmatic crests. It is vascularized by three veins, one dorsal and two ventral. During further specialization with the fusion of the borders of the conduplicate stylodium, the decurrent stigmatic crests are gradually localized at its apex, forming a typical capitate stigma clearly expressed, for example, in Dilleniaceae and in some other families. But even the capitate stigma often has a more or less bilobed character, showing thereby its dual nature.

In spite of the increasing localization of the stigmatic tissue at the apex of the stylodium during evolution, its close communication with the placentae continues to be maintained. This connection is retained on account of the fact that the inner glandular surfaces are modified into a special tissue termed transmitting tissue by Arber (1937). In the most primitive carpels, there is still no differentiation into the strictly stigmatic and transmitting tissues, but with the origin and specialization of the stylodium, the inner transmitting tissue serves as a path for the pollen tube from the localized stigma to the placentae. The ventral surfaces of the primitive conduplicate stylodium may fuse forming a hollow tube that is lined either by a continuous layer of the transmitting tissue or by one or more longitudinal strands of such tissue. But, in most cases, the concrescence is so complete that a solid stylodium is formed without a core of transmitting tissue.

2.6.2. *Evolution of the Gynoecium*

The most archaic taxa of flowering plants are usually characterized by an apocarpous gynoecium. In more primitive apocarpous gynoecia, the carpels are arranged spirally; but, in more advanced gynoecia, the arrangement is cyclic. With the evolution of apocarpous gynoecium the number of carpels decreases and in some extreme cases, such as *Degeneria, Consolida,* or almost the whole family Fabaceae, the gynoecium is monomerous.

Already in the most archaic families, a tendency is observed towards a greater or lesser union of carpels which leads to the formation of the syncarpous (coenocarpous) gynoecium. As a result, forms with more or less syncarpous gynoecia appear even in such families as Magnoliaceae, Annonaceae, Winteraceae, etc. The union of carpels has occurred independently and heterochronously in many different evolutionary lineages and the overwhelming majority of the magnoliophytes has one or another type of syncarpous gynoecium. All types of carpel "from typical conduplicate to extreme involute and from widely open to completely closed" have apparently entered into the formation of synocarpous gynoecia (Eames 1961:233).

But what is the adaptive advantage of syncarpy and why does it dominate over apocarpy? The most plausible and the simplest explanation is based on the principle formulated by Wernham (1913:5) under the name "the tendency to economy in production of reproductive parts." According to Wernham, economy is a guiding principle in the evolution of the flower—including the fusion of the carpels. Both space and material are economized by the fusion of carpels. Stebbins comes to an analogous conclusion: he says that the most likely advantage of the union of carpels may lie in the fact that gynoecium with fused carpels "elaborates a smaller amount of wall tissue compared with ovular tissue than do separate carpels that produce an equivalent seed mass. Hence in terms of the reproductive effort, as defined by Harper and Ogden (1970), a syncarpous gynoecium may be more efficient than an apocarpous one" (1974:295). Besides, fusion of the carpels to one integral structure opened new possibilities for further adaptive evolution. Of course free carpels, especially those in monocarpellate gynoecia, also undergo various adaptive modifications (the best example being the family Fabaceae), but syncarpous gynoecia are much more liable to complex mechanical and other adaptations for dispersal of seeds or the entire fruits. Endress (1982:48) also emphasizes the importance of "increased diversity of dispersal types in the fruiting stage." Another adaptive significance of advanced syncarpy has been proposed by Carr and Carr (1961) and later by Stebbins (1974) and Endress (1982). According to Carr and Carr, if the union of carpels is complete and the stylodia are fused (even the stigmas are separate), there is "a compitum, a connection between the carpels which allows pollen tubes from grains germinating on any stigma or part of the stigma to fertilize ovules belonging to more than one carpel. In multilocular ovaries the compitum characteristically consists of pores, ducts or splits in the septa between loculi, through which the pollen tubes pass to the placentae. In other cases, and typically in unilocular ovaries with parietal placentation, the style may function as a compital mechanism" (1961:255). But, of course, the advantage of compitum is not realized in gynoecia of which the upper parts of the carpels are separate (Stebbins 1974). Therefore, the initial advantage of syncarpy is its parsimonial construction and development.

The syncarpous gynoecium usually originates from a more advanced cyclic apocarpous gynoecium. But syncarpous gynoecia are known also in some archaic taxa with a spiral arrangement of carpels. One of the best examples is the southeast Asiatic genus *Pachylarnax* (Magnoliaceae), a few spirally arranged carpels of which are completely concrescent, forming at maturity a woody loculicidal capsule. No less interesting are two other south Asiatic magnoliaceous genera—*Aromadendron* and *Paramichelia*. Their numerous carpels are concrescent, forming a fleshy berrylike fruit. The carpels are also concrescent in the magnoliaceous genera *Talauma, Kmeria,* and *Tsoongiod-*

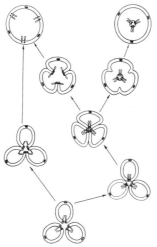

Figure 15. Schematic diagram of the evolution of main types of gynoecium. The apocarpous gynoecium of unsealed carpels (bottom) gives rise to the paracarpous gynoecium (top left) and the syncarpous gynoecium (right); the syncarpous gynoecium in its turn gives rise to the paracarpous (top left) and the lysicarpous (top right) (adapted from A. Takhtajan 1959).

endron. In *Talauma*, the carpels are concrescent at least at the base, the fruiting ones woody, circumscissile. In the genus *Zygogynum* sensu stricto (Winteraceae), the gynoecium is also syncarpous but the arrangement of carpels is cyclic rather than spiral; as is true in other syncarpous members of archaic taxa.

I distinguish three main types of syncarpous gynoecium: eusyncarpous, paracarpous, and lysicarpous (figure 15). The eusyncarpous gynoecium (or syncarpous sensu Troll) denotes a multilocular gynoecium consisting of varying number of coalesced carpels (Grisebach 1854; Hofmeister 1868; Troll 1928, 1957; Takhtajan 1942, 1948; Baum 1949). It emerged independently in many lines of evolution from an apocarpous gynoecium by lateral concrescence of closely connivent carpels. The eusyncarpous gynoecium usually originates from a more advanced cyclic apocarpous gynoecium with more or less sealed carpels, although among Magnoliaceae various stages of the concrescence are observed already within the spiral gynoecium. The union of the carpels occurs either in the course of ontogeny (and in this case, the early stages of development recapitulate apocarpy) or otherwise it is congenital, i.e., the gynoecium develops from the very beginning as an entire structure (Baum 1949). In certain families, various stages of the formation of the eusyncarpous gynoecium may be traced. All the intermediate forms between the typical apocarpous gynoecium and syncarpous type are observed within the families Cunoniaceae, Crassulaceae, Rosaceae, Melanthiaceae, or Arecaceae. The most primitive forms of eusyncarpous gynoecium still have free upper portions of the fertile region of the carpels (as in *Trochodendron* or in many Crassulaceae). With specialization of the eusyncarpous gynoecium, the concrescence extends also to the stylodia, which finally coalesce completely into a style with one apical stigma. All the transitions are observed from completely free stylodia through the stylodia coalesced near the base into a short style with free stylar or stigmatic branches at the top, to perfectly entire styles ending with the apical stigmatic head. The increasing concrescence of the carpels leads also to anatomical changes: with close fusion of carpel margins, the epidermal layers on the surface of contact are lost and the two ventral bundles form a single bundle (Eames 1931).

The paracarpous gynoecium evolved in many lines of both magnoliopsids and liliopsids. It is found much more frequently than the syncarpous gynoecium and is thus the most widespread type. Usually, the paracarpous gynoecium denotes a unilocular gynoecium, consisting of several carpels and having parietal or free-central placentation (Grisebach 1854; Hofmeister 1868; Troll 1928, 1957; Just, in Gundersen 1950; Engler's Syllabus in Melchior, ed. 1964; Ehrendorfer, in Strasburger 1983). But I prefer to limit the concept of paracarpous gynoecium to only the form of unilocular syncarpous gynoecium that has a parietal arrangement of ovules (Takhtajan 1942, 1948, 1959, 1980). A paracarpous gynoecium is characterized by unfolded individual carpels. Their margins are disconnected, while the connection of the borders of the adjoining carpels is maintained (figure 15).

The paracarpous gynoecium usually originates from the apocarpous gynoecium with free, unsealed carpellary margins. Less frequently, it could evolve from the primitive eusyncarpous gynoecium with still distinct and not yet fused carpellary margins. With the evolutionary transformation of the apocarpous gynoecium into the paracarpous, the carpels are unfolded, accompanied by the coalescence of the borders of the adjoining carpels or lateral concrescence. In a case where paracarpous gynoecium evolves from the primitive eusyncarpous gynoecium, only the unfolding of each individual carpel takes place. All the intervening stages of the formation of a paracarpous gynoecium from the eusyncarpous one can be traced, starting from such forms where the lower portion is still eusyncarpous (as in *Parnassia*) to those where the whole gynoecium became paracarpous. The transition of both the apocarpous and eusyncarpous gynoecia into the paracarpous is realized at a relatively early stage of ontogenetic development.

The paracarpous gynoecium appears independently in quite different evolutionary lines and characterizes very many families and entire orders, including two largest families—Asteraceae and Orchidaceae. Structurally, the paracarpous gynoecium is very economical: with a relatively low expenditure of the building materials, it ensures fairly effective protection of the ovules, and the placentae, freed from the spatial restriction of the ventral angle of the folded carpels, gets the possibility of specialization in different ways and often expands, branches, and attains a fairly complex structure (Takhtajan 1964).

The paracarpous gynoecium probably appears very early in the angiosperm evolution. It is already found among Magnoliidae where it is present in the African annonaceous genera *Monodora* and *Isolona* (Deroin 1985), in the whole family of Canellaceae, and in the madagascan winteraceous genus *Takhtajania* (Leroy 1978; Vink 1977). Paracarpous gynoecia originated from primitive apocarpous gynoecia with unsealed carpels. In many families, both apocarpous and paracarpous taxa are observed. Besides the above-mentioned Annonaceae and Winteraceae, we may mention Saururaceae, Cactaceae, and many other families. In the family Saururaceae, the genus *Saururus* has an apocarpous gynoecium and the remaining four genera are paracarpous. In the family Cactaceae, the gynoecium in species of the primitive genus *Pereskia* is semiapocarpous (carpels only weakly united), while in the other genera it is paracarpous. In the archaic monocotyledonous subclass Alismatidae, the families Butomaceae, Limnocharitaceae, and Alismataceae are characterized by apocarpous gynoecium, while in all Hydrocharitaceae it is paracarpous. There is little doubt that paracarpous gynoecium in subclass Dilleniidae originated from the apocarpous Dillenialeslike ancestors.

In many cases, the placentae in the paracarpous gynoecium grow thick, expand, and intrude inside the ovarian cavity where they meet and often coalesce, forming false septa and pseudoaxile placentation, as for example in the family Campanulaceae. Puri (1952) is quite right in inclining to the conviction, that the multilocular character of this type, i.e., which appeared due to the concrescence of the placentae and not carpellary margins, is more common than was earlier thought. It is not difficult in most cases to distinguish the multilocular paracarpous (pseudosyncarpous) gynoecium from the true eusyncarpous gynoecium. The true eusyncarpous gynoecium is usually characterized by a weak development of the purely sutural placentae. Such, for example, is the structure of eusyncarpous gynoecia in *Nepenthes, Nigella,* Hamamelidaceae, *Daphniphyllum, Balanops, Myrotham-*

nus, Buxaceae, many Rosidae, many Liliales, Amaryllidaceae, etc. In several lines of evolution of magnoliopsids, for example in Primulales, the multilocular gynoecium gave rise to a special type of unilocular gynoecium which I named lysicarpous (Takhtajan 1942, 1948, 1959). Like the paracarpous gynoecium, the lysicarpous type is also unilocular but it is characterized by free-central ("columnar") placentation instead of parietal (figure 15). The unilocular ovary in the lysicarpous gynoecium is due to the disappearance of the septa of the multilocular ovary, which takes place either during ontogeny, as in Portulacaceae and some Caryophyllaceae, or during evolution, as in Primulaceae. The united placental areas remain entire and the ovules continue to be perched on them as earlier (for the literature, see Puri 1952). Thus, the united placentae are transformed into a column freely rising at the center of the ovarian cavity and usually not reaching the top of the ovary.

Certainly all three types of the syncarpous gynoecium are connected by transitions, and in many cases the structure of gynoecium has an intermediate character between them. But I cannot agree with Parkin (1955), who on the basis of absence of "hard and fast" distinctions between the axile and parietal placentation, opposes the classification proposed by Troll. It is difficult to blame Troll for extra details in his classification, as his paracarpous type is, on the contrary, heterogeneous and requires subdivision into two separate types—the paracarpous (in the narrow sense of the term) and the lysicarpous.

Specialization of the syncarpous gynoecium as well as that of the apocarpous is usually (but not always) accompanied by greater or lesser reduction in the number of carpels and, in most cases, also by the number of ovules. An extreme form of reduction in the number of carpels in the syncarpous gynoecium is the so-called pseudomonomerous gynoecium (Eckardt 1937, 1938), where only one of the carpels is fertile. The pseudomonomerous gynoecium results from a sharply expressed dissimilar development of the carpels when one carpel dominates and attains full development and the rest are undeveloped and remain sterile and rudimentary. The sterile carpels (or carpel, if the gynoecium is dimerous) in the pseudomonomerous gynoecium often attain such a degree of reduction that their presence can be detected only through an anatomical study of the vascular system and ontogeny. Although externally the pseudomonomerous gynoecia has the appearance of a solitary carpel, at least two carpels take part in its formation. The pseudomonomerous gynoecia of the dimerous type characterizes, for example, *Casuarina, Eucommia,* Ulmaceae, Moraceae, Cannabaceae, Cecropiaceae, Urticaceae, *Barbeya,* a majority of Thymelaeaceae, *Leitneria,* Mastixiaceae, *Aucuba, Garrya, Hippuris,* Globulariaceae, etc., and those of trimerous and polymerous types characterize a majority of Piperaceae, and Chrysobalanaceae, Valerianaceae, and some other families.

2.6.3. *Evolution of Placentation*

The main directions of evolution of the gynoecium determine the main trends of evolution of placentation.

We find two main types of placentation in flowering plants: laminar and submarginal (sutural). In the first case, the ovules sit on the inner surface of the carpellary lamina; in the second, along the sutures very near the margins of the carpels. Submarginal placentation is not marginal in the strict sense. Investigations of the vascular anatomy and the early stages of development of the gynoecium in various taxa, particularly in the archaic forms, clearly indicate that in submarginal attachment, the ovules sit not exactly at the margins but on the adaxial side near the margins, i.e., they are submarginal (Tepfer 1953; Takhtajan 1959:102; Eames 1961:208). Superficial impression about the involuteness of marginal borders and marginal character of placentation in a number of flowering plants is a result of the evolutionary change of carpel ontogeny. Therefore, the submarginal placentation may be considered rather as a product of laminar placentation.

The types of placentation may be classified as follows:

A. Laminar placentation. The ovules occupy the side portions of the inner face of the carpel or are scattered over almost the entire surface, rarely occupy only its back side.
   1. Laminar-lateral placentation. The ovules occupy the side portions of the inner face between the median and lateral veins. Examples: *Tasmannia, Degeneria.*
   2. Laminar-diffuse or scattered placentation. The ovules are scattered over almost the entire inner surface of the carpels. Examples: Nymphaeaceae, Butomaceae, Limnocharitaceae, Hydrocharitaceae.
   3. Laminar-dorsal placentation. The ovules are attached pseudomedially, occupying the back of the carpel. Examples: *Nelumbo, Ceratophyllum.*
B. Submarginal (sutural) placentation. The ovules occupy morphologically sutural areas of the carpel.
   4. Axile placentation. The ovules are attached along the suture of the carpel very near the approximated or fused margins of the carpel in an apocarpous or eusyncarpous gynoecium. Examples: Ranunculaceae, Dilleniaceae, Rosaceae, Liliaceae.
   5. Parietal placentation. The ovules are situated along the sutures in a paracarpous gynoecium or on the intrusive placentae which in turn are attached to the sutures. Examples: Violales, Capparales, Orchidales, Juncales.
   6. Free-central or columnar placentation. The ovules are situated along the central column of the lysicarpous gynoecium. Examples: Portulacaceae, Myrsinaceae, Primuluceae.

The most primitive type of placentation is laminar-lateral (Takhtajan 1942, 1948, 1959, 1980; Eames 1961; Stebbins 1974). It characterizes such archaic plants as *Degeneria* and *Tasmania* and certain species of the genus *Zygogynum,* specially *Z. archboldianum* (*Bubbia archboldiana*) (Bailey and Smith 1942; Bailey and Nast 1943; Swamy 1949; Bailey and Swamy 1951). The ovules of these plants are rather far away from carpellary margins and are arranged in the space between the median and lateral veins. The ovules are vascularized in part by extensions of the branches of the lateral systems, in part by extensions of the veinlets of the median and lateral branches. Such an arrangement of ovules is most probably very near to the initial one in the evolution of angiosperm placentation.

The laminar-diffuse (scattered) placentation is very close to the laminar-lateral type and essentially is a variation of it. We can see the laminar-diffuse placentation already in some archaic taxa.

In the Himalayan-Chinese genus *Decaisnea,* the most archaic genus in the family Lardizabalaceae, as well as in the Chinese genus *Sinofranchetia,* the ovules are arranged in two rows, i.e., the placentation here is still of laminar-lateral type. But in the other genera of this family the ovules are already arranged in several longitudinal rows, i.e., more or less "diffuse." Rarely are the ovules few or solitary.

In Nymphaeaceae, Barclayaceae, Butomaceae, Limnocharitaceae, and Hydrocharitaceae, the laminar-lateral placentation also acquired a diffuse character. According to some authors, the diffuse placentation is the most primitive type of ovule position (Kozo-Poljanski 1922, 1928; Zazhurilo and Kuznetsova 1939; Joshi 1947; Gundersen 1950). But the diffuse type of laminar placentation is found only in the carpels with one leaf trace and is unknown in more primitive three-trace carpels. Besides, carpels with diffuse placentation are generally more specialized than those with the laminar-lateral placentation. There is strong evidence that laminar-diffuse placentation evolved from still more archaic laminar-lateral placentation (Takhtajan 1948, 1959:102). Parkin (1955) is also inclined to conclude that the laminar placentation of the type of Nymphaeaceae and Butomaceae is secondary.

With the transition from laminar placentation into the strictly submarginal, a progressive reduction generally takes place in the number of ovules in each carpel, which can be noted already within the family Winteraceae. The decrease of the quantity of ovules is related to the improvement of their protection and to the perfection of the disseminution mechanism. This evolutionary trend culminates in a carpel with a single ovule. In this respect, one must note a striking resemblance to the mammals, in which too, evolution was directed towards a reduction in the number of embryos. In both cases the economy of material is achieved by a better "care about the progeny."

Certain groups of magnoliophytes with apocarpous gynoecia but with a reduced number of ovules show a special type of placentation, at times called dorsal or "median." It is found in *Brasenia, Cabomba, Ceratophyllum, Nelumbo,* Potamogetonaceae, and some others. But as the vascular anatomy of the carpels of these plants indicates, such placentation is not "median" in the strict sense of the term (Saunders 1936; Puri 1952; Eames 1961, etc.). Thus, in *Cabomba* and *Brasenia,* the ovules are situated actually between the lateral veins and the median vein, and the vascular bundles of the funicles are related both to the lateral and median veins. The vascular system of the carpel and placentation in these two genera clearly shows that their dorsal placentation evolved from the laminar-lateral.

Submarginal (sutural) placentation evolved from laminar placentation. This is the most widespread type of placentation in flowering plants, which is already found in a majority of taxa with apocarpous gynoecium including Magnoliaceae, Annonaceae, and Ranunculaceae, and characterizes the carpels of both the follicle and achenium type. In the archaic forms, the carpels of which are still follicles, the ovules are situated in the rows along the suture with one row on each side. Such genera as *Trollius* or *Helleborus* still have typical multiovulate carpels with a primitive vascular system. But already in the genus *Trollius*, the commencement in the reduction of the number of carpels can be observed: the upper ovules in the carpel have vanished here but their vascular traces are still present. The next step in the reduction series can be observed in *Aquilegia*, where not only the upper ovules themselves but also their vascular traces have been lost. In the genus *Ranunculus*, with its specialized carpels of the achenium type, only one lower ovule develops. A similar reduction series can be found in Rosaceae (Chute 1930). The greatest variety of forms of submarginal placentation can be observed in eusyncarpous gynoecia. The main and initial type is the axile placentation. Emerging already in the free carpel of an apocarpous gynoecium, the axile placentation undergoes further modifications in the eusyncarpous gynoecium. If, in the separate carpels of the apocarpous gynoecium, its two lateral veins gradually converge and then merge together; then, in the syncarpous gynoecium, the sutural bundles of the neighbouring carpels coalesce, as well. This fusion of the sutural bundles is usually accompanied by a reduction in the number of ovules. In some cases, due to ontogenetic shifts, the axile placentation appears to be diffuse, as for example in *Mesembryanthemum* and *Punica granatum*. In such cases, the study of ontogeny indicates that the ovules are situated on the axile placentae but due to differential growth of the gynoecium and change in the position of the ovules the placentation appears to be diffuse (Buxbaum 1951; Puri 1952; Sinha and Joshi 1959).

Parietal placentation appears independently and heterochronously in many lines of evolution of both the magnoliopsids and liliopsids. This indicates that it has a definite biological advantage. In addition to the constructive simplification of the paracarpous gynoecium (unilocular design instead of the multil-ocular), the placentae get a wider possibility for specialization in various directions.

Lastly, parietal and possibly in some cases also axile placentation give rise to a free-central or columnar placentation. The origin of the free-central placentation was for long an object of dispute. In the first half of the last century, most botanists considered that the central column is simply a result of the elongation of the receptacular axis. But starting from the classical investigations of Van Tieghem (1868), it became clear that the central column is a result of the degeneration of the septa with the placentae preserved in the center. Lister (1883) arrived at a similar conclusion on the basis of a study of a number of representatives of the family Caryophyllaceae. Subsequent investigations of ontogeny and comparative vascular anatomy of the various types of free-central placentation finally confirmed the carpellary nature of the central column (Bancroft 1935; Dickson 1936; Douglas 1936; Laubengayer 1937; Wilson and Just 1939; Thompson 1942; Joshi 1947; Takhtajan 1948; Puri 1951, 1952; Eames 1951, 1961; etc.).

2.6.4 Origin of the Inferior Ovary

In archaic flowering plants, the gynoecium is still free and does not yet coalesce with the floral parts surrounding it. But in many lines of evolution of magnoliophytes, the gynoecium more or less coalesces with the neighboring parts of the flower and as a result the so-called "inferior ovary"* is formed. Grant (1950) concludes, that the inferior ovary originated as a protective

---

*The term "inferior ovary" is unfortunate. Morphologically, the ovary cannot be "inferior"—it is always "superior," as the carpels are always situated morphologically above the perianth and the stamens. Therefore, it would have been advatageous to replace the terms "superior" and "inferior" ovary by more exact ones—"free" and "adnate" ovary.

adaptation against the insects and birds, pollinating the flowers. But I agree with Stebbins (1974:305), that different trends of epigyny may well have been guided by entirely different selective pressures such as protection of developing flowers against biting insects and from environmental shocks and more rapid development of the flower and maturation of the fruit.

"The morphology of the inferior ovary has doubtless been discussed more extensively than that of any other part of the plant body," says Eames (1961:245). Some authors think that the inferior ovary originates through the invagination of the floral receptacle, which surrounds the gynoecium and adnates to it (the axillary or receptacular theory). Others think that it is formed by the adnation of the bases of the outer floral whorls to the gynoecium (phyllome or appendicular theory). Still others, like Gertrude Douglas (1944, 1957) and Eames (1961), presume both the axial and appendicular modes of origin of the inferior ovary. But, according to Eames, nearly all inferior ovaries seem to have been formed by the fusion of adjacent floral organs.

Numerous data on vascular anatomy of the flower lead to the conclusion that, in an overwhelming majority, the inferior ovary resulted from the coalescence of the gynoecium with the floral tube, i.e., it is of phyllome origin. It has been firmly established by the classic works of Van Tieghem (1868, 1875) and especially by more recent works on floral anatomy, that in a majority of angiosperm families, the floral tube* surrounding the inferior ovary is a product of the fusion of the bases of the sepals, petals, and stamens and consequently has a phyllome character. The phyllome nature of the floral tube can be detected by the com-

*The term "hypanthium," frequently used in literature instead of the term "floral tube," is wrong because the tube is not below the flower (Jackson 1934). Parkin (1955:54) also thinks, that "hypanthium is a superfluous botanical term." The term "hypanthium" might have been kept only for the receptacular cup of the type *Rosa* and *Calycanthus*.

parative study of flower. Van Tieghem already established how the "appendicular" tissues of the inferior ovary can be distinguished from the receptacular ones by the arrangement of the vascular system. Anatomical studies show that, due to fusion of the members of a particular whorl among themselves (cohesion) and the fusion of different whorls (adhesion), their vascular bundles also fused to a greater or lesser degree. At the first stages of fusion of the floral parts, their vascular bundles still remain free but, subsequently, they gradually converge and finally may fuse together. The study of the vascular bundles and the various stages of their fusion in a series of related forms provides a key to the history of fusion of the floral parts. On the basis of his own studies and those of his students, Eames (1961) came to the conclusion that fusion of adjacent organs is a common mode of the origin of inferior ovary (see especially Douglas 1957, where the basic literature is cited).

But it is becoming clear now, that in certain taxa, the inferior ovary has—all the same—a receptacular and not a phyllome origin. Neither at the time of Van Tieghem and Henslow (1891) nor even in the first decades of this century, was there found any anatomically substantiated and undoubted example of the inferior ovary of receptacular origin. Only in 1942 was an extremely interesting example of the true receptacular inferior ovary first described in *Darbya*, from the family Santalaceae (Smith and Smith 1942). The receptacular inferior ovary presumably characterizes all the members of Santalaceae and also the related and more specialized families Misodendraceae and Loranthaceae *sensu lato* (Smith and Smith 1942; Schaeppi and Steindl 1942, 1945). The vascular anatomy clearly indicates the presence of invagination of the stelar apex and sinking of the gynoecium in the receptacle. The receptacular inferior ovary is characterized by the presence of recurrent vascular bundles, which after reaching the upper portion of the ovary, turn sharply inward and downward and bend towards the morphological apex of the receptacle occupying the lowermost position at the bottom of the cup-shaped depression. In these recurrent bundles, the phloem is situated adaxialy (is inwardly directed) and the xylem is abaxial, i.e., these bundles are inverted. This tendency towards the invagination of the receptacle and recurrection of its vascular system may be observed in the members of the family Olacaceae—the most archaic in the order Santalales.

The receptacular origin of the inferior ovary has been established also in the families Cactaceae (Buxbaum 1944; Zamyatnin 1951; Tiagi 1955; Boke 1966) and Aizoaceae (Zamyatnin 1951). In both, the receptacular origin is confirmed on the basis of anatomical analysis.

Thus, both the phyllome and receptacular inferior ovaries evolve independently in different lines of evolution of flowering plants.

2.7. The Evolution of Inflorescences

According to Braun (1875), Hallier (1901, 1902, 1912), Parkin (1914), and many others, the initial form of arrangement of flowers is their solitary disposition at the end of a leafy branch. But among living flowering plants, solitary flowers—both terminal and axillary—most probably represent the surviving members of reduced inflorescences (Eames 1961; Foster and Gifford 1974; Stebbins 1974; Takhtajan 1980). In the Winteraceae, for example, "the terminal single flowers of certain species of *Zygogynum* must represent the end of a reduction series" (Bailey and Nast 1945a:38). Usually the secondary character of solitary flowers can be explained on the basis of morphological analysis and comparison with related forms.

The flowers of the earliest flowering plants were probably already aggregated into a very primitive inflorescence. The aggregation of flowers must have facilitated their pollination: each pollinator could visit many more flowers assembled in an inflorescence than the solitary flowers per unit of time. The successive opening of the flowers in an inflorescence is also of a great biological advantage against the momentary opening of a single flower. The inflorescence also helps in perfecting cross pollination.

In less modified inflorescences, each of the successive flowers is subtended either by only somewhat diminished and slightly changed vegetative leaves (leafy inflorescences) or by a bract which represents much reduced and simplified leaf (bracteose inflorescences). There are also every kind of intermediate types.

Beginning with authors of the last century, the various types of inflorescences are usually divided into two major categories —cymose, determinate, centrifugal, or closed, and racemose (botryose), indeterminate, centripetal, or open. According to Weberling, these two types of inflorescence correspond to Troll's (1964) monotelic and polytelic inflorescences, although "with some essential reservations" (1965:216). In the cymose inflorescence, the apex of the inflorescence axis ends with a terminal flower and it is therefore characterized by the limitation of the apical growth of the main axis. The terminal flower of the main axis typically opens first and the order of flowering is therefore usually in downward (basipetal) sequence. In the racemose inflorescence, there is no terminal flower at the apex of the main axis. Therefore, there is no limitation of the apical growth of the primary axis. The order of flowering in racemose inflorescence is typically in upward (acropetal) sequence.

But since Eichler (1875), it is well known that the difference between cymose and racemose inflorescences is not clear cut. While the more primitive and more simple forms of two basic types of inflorescences are usually easily distinguished, their most specialized and particulary reduced forms are often indistinguishable or almost indistinguishable from each other. "Since many racemose inflorescences have a terminal flower and many cymose inflorescences have none, the presence or absence of a terminal flower on the main axis is not a safe criterion for the separation of the two groups," states Rickett (1955:425). There are many intermediate forms, especially racemose inflorescences which have a terminal flower, as in *Berberis, Juglans, Clethra, Pyrola, Mitella,* certain species of *Ribes* (e.g., *R. aureus*), *Rubus, Prunus serotina, Robinia pseudacacia,* various species of *Diervilla, Digitalis, Campanula, Convallaria, Camassia,* etc. On the other hand, some cymose forms such as *Agrimonia eupatoria* do not end in a terminal flower.

There are also intermediate forms in the order of flowering. In the dipsacaceous genus *Cephalaria,* the inflorescence is not always strictly cymose and the outermost (lowermost) row of the flowers may open immediately after the terminal flower (Cronquist 1968; and also my own observations). "The order of flowering is usually but not necessarily acropetal in a raceme. The order of flowering is typically basipetal or centrifugal in a single three-flowered dichasium, but not in an inflorescence made up of many such units. In many compound inflorescences no regular sequence of flowering is to be seen" (Rickett 1955:425). There are, indeed, many examples where there seems to be no obvious sequence of flowering. Rickett mentions *Philadelphus coronarius, Rubus allegheniensis,* and *Sambucus canadensis.*

But the number of exceptions and intermediate forms between cymose and racemose inflorescences is not so very great. I agree with Stebbins (1974:264), who says: "Inflorescences that belong to one or other of these categories far outnumber those that are intermediate. Moreover, in very many plant families, probably the majority of them, inflorescences are either exclusively determinate or indeterminate as to their main axis." There will be even fewer intermediate forms if—as the main criterion—we accept, not the order of flowering, but rather the presence or absence of the terminal flower at the end of the main axis (monotelic or polytelic!).

Of two basic groups of inflorescences, the cymose inflorescence is more primitive and the racemose inflorescence is derived (Parkin 1914; and many others). Weberling (1965:220) also comes to the conclusion that in general the polytelic type is more highly evolved than and perhaps derived from the monotelic type. "Even within several polytelic families it should be noted that the more primitive genera still have monotelic inflorescences," says Weberling.

The most primitive form of inflorescence might be either a simple leafy cymose "panicle" or simple leafy dichasium. Already Nägeli (1884) concluded that both cymose and racemose inflorescences could be derived from a "panicle." Later, Pilger (1922) and Weberling (1988) also accepted the panicle as the most generalized type (but not necessarily as a primitive type, according to Weberling). According to Zimmermann (1935, 1965), who accepted the derivation of the racemose type from the cymose one, the most primitive type of inflorescence is the "cymöse Rispe," which he considers as an "phylogenetische Urform." Von Denffer in Strasburger's *Lehrbuch der Botanik* (1983) also accepted as the most primitive type of inflorescence "die geschlossene Rispe" (closed panicle), e.g., panicle of *Vitis*. From this primitive panicle with alternate branches, he derives "decussierten geschlossenen Rispen," exemplified by *Syringa*. Although from the cymose panicle it is easy to derive typical dichasial inflorescences, it seems more likely that the original condition was the leafy dichasium, probably a simple one.

From the simple leafy dichasium originated various kinds of cymose inflorescences as well as solitary flowers (figure 16). By complete suppression of lateral flowers in a simple dichasium, solitary flowers evolved. A single terminal flower of *Zygogynum vieillardii, Liriodendron tulipifera,* or *Paeonia suffruticosa* probably developed by complete suppression of lateral flowers.

By means of repeated branching of the successive orders of branches, a simple dichasium gives rise to a more or less pani-

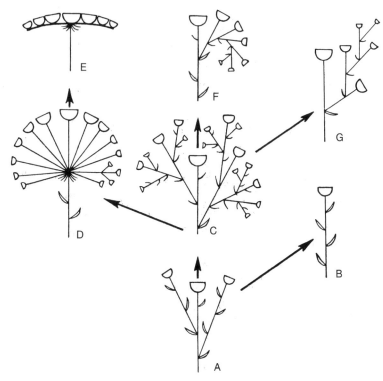

Figure 16. Diagram of the evolution of main types of cymose inflorescences. A, simple leafy cyme; B, solitary terminal flower; C, compound cyme; D, umbelliform cyme; E, capitate cyme; F, helicoid cyme; G, scorpioid cyme. (adapted from A. Takhtajan 1959).

culate compound dichasium. Thus emerges, for example, the loose dichasial inflorescence of some species of *Ranunculus* and *Potentilla* and many Caryophyllaceae. Some species of *Ranunculus* even have floral branches of the fifth order. As a result of the complete suppression of all lateral branches in compound dichasia, solitary terminal flowers evolved.

Through an abbreviation of the internodes or undevelopment of certain flowers, various modifications of the loose dichasium emerge (corymbose dichasium, umbelliform dichasium, capitate dichasium, etc.).

With evolution, the compound dichasia undergo drastic changes, often making it difficult to find out their true morphological nature. The most interesting in this respect are the capitate dichasial inflorescences of the Dipsacaceae, certain Rubiaceae (e.g., *Cephaëlis ipecacuanha*) and Valerianaceae and so-called syconium of certain Moraceae. In the family Rubiaceae, evolution starts from the simple dichasia, which change into compound dichasia, which in their turn give rise to dense spherical heads by the aggregation of dichasia as in *Uncaria* and some other related tropical genera or in the genus *Morinda*. In these heads, the ovaries often coalesce completely. Capitate dichasial inflorescences are found also in the valerianaceous genera *Nardostachys* and *Plectritis*.

The compound dichasia undergo much larger modifications in the representatives of Urticaceae and Moraceae (see Berg 1990). The most remarkable are the highly specialized inflorescences of *Dorstenia* and especially of *Ficus,* which are very different from any usual type of dichasial inflorescence. But already Eichler (1878) considered the inflorescence of *Ficus* dichasial, where all the axes are fused into a general mass.

The evolution of dichasial inflorescences is also very complicated in the families Betulaceae, Fagaceae, and Leitneriaceae, where they are very specialized and extremely modified (see Abbe 1974).

From the dichasial inflorescences evolved both kind of monochasial inflorescences—simple and compound. By complete suppression of one of the two lateral flowers of the simple dichasium, there evolved the monochasium, such as, for example, in *Anemonella thalictroides*. In certain genera or even families, by suppression of one of the two branches of ramification of the compound dichasium there originated a compound monochasium. In some other cases, by complete suppression of the single lateral flower in monochasium, a solitary flower evolved.

The compound monochasia are of two main types. When each new lateral axis appears successively on one and the same side of the sympodial main axis, with the result that it is curved or circinately rolled, the inflorescence is called helicoid cyme.

Another form of compound monochasium is the scorpioid cyme, which is formed by repeated monochasial branching in which successive branches arise alternately on opposite sides of the sympodial axis and therefore flowers are following a descending spiral.

In its turn, compound monochasium can undergo certain modifications, which sometimes completely change its general appearance. The most remarkable modification of the compound monochasium is an umbelliform monochasium, which originates by means of the abbreviation of the main axis. As a result, the branches of the cyme become so contracted that they appear to arise from almost the same point, and, therefore, this inflorescence imitates an umbel. The umbelliform monochasium consists of several monochasia with contracted internodes.

Another modification of the compound monochasium, which is characteristic for certain lilopsids, is rhipidium. The rhipidium is a flattened in one plane, scorpioid, fan-shaped cyme in which lateral flowers develop on opposite sides of the main axis alternately. Rhipidia are found in many Iridaceae.

The racemose inflorescence evolves from the cymose inflorescence, particularly from the cymose panicle. According to Parkin (1914:516), "a panicle in its simple form is nothing but a cyme, and in its more complicated form a transition to a true raceme." The transition from a simple panicle to a true raceme is gradual. The raceme is a more or less elongated—and in its typical form, polytelic—inflorescence, with the flowers on the lateral pedicels commonly arising in the axils of bracts. In typical racemes, flowering is acropetal.

The transitions from the cymose panicle to the raceme (figure 17) are found in many relatively less advanced families—for example, in Ranunculaceae. Thus, in *Consolida divaricata* and *Aconitella hohenackeri,* the inflorescence is a cymose monotelic panicle. But the inflorescences of many other species of these genera represent an intermediate stage. Finally, the apical flower vanishes completely and the inflorescences become a typical raceme. Such a true raceme is characteristic, e.g., for *Consolida orientalis, C. camptocarpa, C. leptocarpa, C. persica,* and for the majority of species of the genus *Delphinium.* At first sight, it appears that there is an apical flower in the inflorescence of *Consolida orientalis* and some other species of the genus, but in reality, the uppermost flower of the inflorescence is morphologically a lateral flower which arises in the axile of the bract and usually bears a pair of bracteoles on its pedicel. The apical flower in *Consolida orientalis* and related species is completely obliterated and there are no remnants at all. This same trend from the cymose to racemose inflorescence may be observed within the genus *Aconitum.* In some species of *Aconitum,* as, for example, in *A. arcuatum* and *A. anthoroideum,* the apical flower is preserved, while in the other species, such as *A. anthora,* the apical flower does not develop at all (Takhtajan 1948).

The family Fumariaceae is also a good illustration of the passage from panicle to raceme (Parkin 1914). *Dicentra formasa* possesses a branched panicle of four or five lateral branches. "The terminal flower is usually the first to bloom, occasionally the last; afterwards the order of the expansion is inclined to be acropetal. The lateral axes branch dichasially producing tertiary flowers to some extent, and even flower-buds to the fourth degree may be formed" (Parkin 1914:540). Although the terminal flower of *D. formosa* is usually the first, it is already not a typical cymose inflorescence. But in *D. spectabilis,* the inflorescence is already a true raceme. "The uppermost part of the inflorescence is occupied by a few rudimentary buds which do

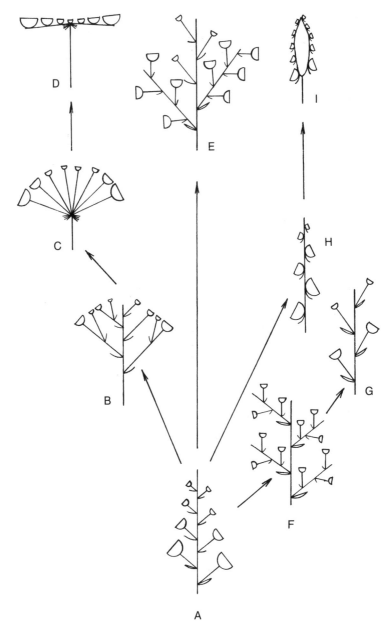

Figure 17. Diagram of the evolution of main types of racemose inflorescences. A, raceme; B, corymb; C, umbel; D, capitulum; E, compound raceme; F, axillary racemes; G, solitary axillary flowers; H, spike; I, spadix (adapted from A. Takhtajan 1959).

not expand. The lateral floral axes bear only minute bracteoles, no tertiary buds being produced" (ibid.). According to the same author, "in *Corydalis lutea* and other species of this genus the raceme is still more pronounced in that it ends in a minute filament, the remnant of the arrested apical part of the inflorescence, and the pedicels are naked, the bracteoles having been wholly suppressed" (Parkin 1914:540). The genus *Campanula* also "illustrates remarkably well the passage from a lax, cymose, panicled inflorescence to a definite racemose one, usually of the spicate type" (Parkin 1914:541).

By shortening of the pedicels of the lateral flowers the raceme often gives rise to the spike (e.g., *Tetracentron, Plantago, Orobanche*). In *Rhodoleia, Trifolium, Campanula glomerata*, and some other plants, a peculiar capitate raceme or spike is formed which should not be confused with the true cymose head.

The thickening of the spike axis, turning fleshy, results in the spadix which characterizes the Araceae. But it should not be confused with the false spadix of *Typha* or *Zea mays*, which are compound inflorescences.

Another modification of the spike is the ament or catkin distinguished by a slender, flexuous, often pendulous main axis bearing unisexual, apetalous flowers and, finally, deciduous as a whole. The inflorescences of the Salicaceae are typical aments, while in Betulaceae and Fagaceae they are catkinlike dichasial inflorescences.

The raceme gives rise also to the corymb, a special kind of shortened flat-topped convex or concave raceme with all the pedicellate flowers arranged on more or less horizontal plane. Such, for example, are the inflorescences of various species of *Rubus, Viburnum, Hydrangea, Valeriana, Sambucus*, Brassicaceae, etc. In some cases, as in *Filipendula vulgaris*, the branches are almost erect and the lower branches overtop the upper ones. For such a modified corymb, there is a special term: anthela.

The corymb in its turn gives rise to the racemose umbel

(umbellum, *sensu stricto*) which is a flat-topped or convex inflorescence with the pedicels arising at a common point. Like the corymb, the true racemose umbel is a contracted raceme. The transition from the corymb to the umbel is well marked, for example, in the species of *Siphocampylus* (Lobeliaceae). In *S. corymbiferus,* the inflorescence is a corymb; while in *S. lantanifolius,* it is already an umbel. The umbel characterizes Primulaceae, many Araliaceae, a majority of Apiaceae, and others. But the inflorescences of Geraniaceae, Asclepiadaceae, Alliaceae, and Amaryllidaceae are not true umbels, but umbelliform dichasial inflorescences.

The umbel gives rise to a still more specialized form of racemose inflorescence termed capitulum or head. The capitulum is a group of sessile or subsessile flowers on a compound receptacle or torus. The modification of the umbel consists here in shortening of peduncles which are hardly noticeable or almost not noticeable. The main axis is thickened and becomes flattened, or more or less spherical, or conical. The capitulum is surrounded by one or more series of sterile bracts forming the involucre. The capitulum characterizes certain Apiaceae as *Eryngium* and *Sanicula,* Calyceraceae, and Asteraceae. In all these plants, the capitulum did not arise from the raceme as the capitate inflorescence of clover but from the very umbel (Small 1919, Kozo-Poljanski 1923). According to Stebbins, in Apiaceae and Asteraceae "the compressed inflorescence is clearly derived from an indeterminate condition" (1974:267). Thus, the capitulum of Asteraceae belongs to the racemose inflorescences and originated in a completely different manner than the cymose inflorescence of Dipsacaceae (Wernham 1913; Philipson 1953).

In addition to the purely cymose and purely racemose inflorescences, the mixed forms are also found bearing the traits of both types. The most characteristic type of mixed inflorescence is a more or less ovoid or ellipsoid paniculate inflorescence known under the name of "thyrse" or "thyrsus." Its main axis is monopodial but the lateral axes are cymose. The inflorescence of *Syringa*, for example, belongs to this type. Čelakovsky (1892) presumed the thyrsoid type as the most primitive type of inflorescences. The thyrsoid inflorescences are fairly diverse and are found in widely separated families.

Such are the basic types of inflorescences and main trends of their evolution. In reality, there are many more types of inflorescences and the pattern of their evolution is much more complicated. This diversity of the types of inflorescences is strengthened by the presence of different and sometimes very complex combinations of their basic types. Examples of such secondary or composite inflorescences (inflorescentiae compositae) are compound umbels of Apiaceae or catkinlike compound inflorescences of *Betula, Alnus,* or *Corylus*. Secondary inflorescences are either homogeneous or heterogeneous. In homogeneous secondary inflorescences, the arrangement of the primary inflorescences repeats the arrangement of the flowers in the latter. For example, secondary umbel may consist of primary or partial umbels, as in Apiaceae, or secondary spike may consist of primary spikes, as in many Poaceae. On the other hand, in heterogeneous secondary inflorescences, the arrangement of flowers in partial inflorescences is very different from arrangement in the former. Thus, in Asteraceae, the racemose capitula are typically arranged in cymose secondary inflorescence.

It is most interesting that frequently the ways and trends of evolution of secondary inflorescences repeat those of the primary inflorescences. In many cases, the secondary inflorescences imitate the architecture of the primary one. Such are, for example, the catkinlike inflorescences of Betulaceae, which are so similar to aments of *Salix*. Even more remarkable are the secondary capitula of some Asteraceae, for example those of *Echinops,* which are externally almost indistinguishable from the simple (elementary) capitula. It is also interesting that there is a remarkable parallelism in evolution of composite and elementary capitula of Asteraceae.

References

Abbe E. C. 1974. Flowers and inflorescences of the "Amentiferae." Bot. Rev. 40(2):159–261.

Arber A. 1933. Floral anatomy and its morphological interpretation. New Phytol. 32:231–242.

Arber A. 1937. The interpretation of the flower: a study of some aspects of morphological thought. Biol. Rev. 12:157–184.

Arber A. 1942. Studies in floral structure. VII. On the gynaeceum of *Reseda*, with a consideration of paracarpy. Ann. Bot. 6:43–48.

Arber E. A. N. 1899. Relationships of the indefinite inflorescences. J. Bot. (London) 37:160–167.

Arber E. A. N. and J. Parkin. 1907. On the origin of the angiosperms. J. Linn. Soc. Bot. 38:29–80.

Bailey I. W. and C. G. Nast. 1943. The comparative morphology of the Winteraceae. II. Carpels. J. Arnold Arbor. 24:478–481.

Bailey I. W. and C. G. Nast. 1945a. The comparative morphology of the Winteraceae. VII. Summary and conclusions. J. Arnold Arbor. 26:37–47.

Bailey I. W. and C. G. Nast. 1945b. Morphology and relationships of *Trochodendron* and *Tetracentron*. I. Stem, root, and leaf. J. Arnold Arbor. 26:143–154.

Bailey I. W. and A. C. Smith. 1942. Degeneriaceae, a new family of flowering plants from Fiji. J. Arnold Arbor. 23:356–365.

Bailey I. W. and B.G. L. Swamy. 1948. *Amborella trichopoda* Baill., a new morphological type of vesselless dicotyledon. J. Arnold Arbor. 29:245–253.

Bailey I. W. and B. G. L. Swamy. 1951. The conduplicate carpel of dicotyledons and its initial trends of specialization. Amer. J. Bot. 38:373–379.

Bancroft H. 1935. A review of researches concerning floral morphology. Bot. Rev. 1:77–99.

Baum H. 1949. Der einheitliche Bauplan der Angiospermengynözeen und die Homologie ihrer fertilen Abschnitte. Öst. Bot. Z. 96:64–82.

Berg C. C. 1990. Systematics and Phylogeny of the Urticales. In P. R. Crane and S. Blackmore, eds., evolution, systematics, and fossil history of the Hamameidae 2:193–220. Oxford.

Bierhorst D. W. 1971. Morphology of vascular plants. New York and London.

Boke N. H. 1966. Ontogeny and structure of the flower and fruit of *Pereskia aculeata*. Amer. J. Bot. 53:534–542.

Braun A. 1876 (1875). Die Frage nach der Gymnospermie der Cycadeen erläutert durch die Stellung dieser Familie im Stufengang des Gewächsreichs. Monatsber. Preuss. Akad. Berlin:241–267, 289–377.

Brown R. 1840. On the relative position of the division of stigma and parietal placentae. In miscellaneous botanical works 1:553–563. London.

Buxbaum F. 1944. Untersuchungen zur Morphologie der Kakteenblüte. I. Das Gynoecium. Bot. Arch. 45:190–247.

Buxbaum F. 1951. Grundlagen und Methoden einer Erneurung der Systematik der höheren Pflanzen. Die Forderung dynamischer Systematik im Bereiche der Blütenpflanzen. Vienna.

Canright J. E. 1952. The comparative morphology and relationships of the Magnoliaceae. I. Trends of specialization in the stamens. Amer. J. Bot. 39(7):484–497.

Carlquist S. 1969 (1970). Toward acceptable evolutionary interpretations of floral anatomy. Phytomorphology 19:332–362.

Carr S. G. M. and D. J. Carr. 1961. The functional significance of syncarpy. Phytomorphology 11:249–256.

Čelakovsky L. 1875. Uber den "eingeschalten" epipetalen Staubgefässkzeis. Flora 33:481–489, 497–504, 513–524.

Čelakovsky L. 1876 (1877). Vergleichende Darstellung der Placenten in der Fruchtknoten der Phanerogamen (*Scrophularia, Anagallis*). Abhandl. Königl. Böhm. Ges. Wiss., 6th ser., vol. 8:1–74.

Čelakovsky L. 1893. Gedanken zu einer zeitgemässen Reform der Theorie der Blütenstände. Bot. Jahrb. Syst. 16:33–51.

Čelakovsky L. 1894. Das Reduktiongesetz der blüten. Sitzungsber. Böhm. Ges. Wiss. 3:1–136.

Čelakovsky L. 1896, 1900. Über den phylogenetischen Entwiklungsgang der Blüte und über den Ursprung der Blumenkrone 1, 2. Vestnik Královske Ceske Spolec. Nauk, trida mat. prir. 11:1–91; Sitzungsber Königl. Böhm. Ges. Wiss., Cl. 2, 3:1–223.

Chupov V. S. 1987. Some features of the evolution of stamen and perianth parts in angiosperms. Bot. Zhurn. (Leningrad) 71(3):323–333. (In Russian.)

Chute H. M. 1930. The morphology and anatomy of the achene. Amer. J. Bot. 17:703–723.

Corner E. J. 1946. Centrifugal stamens. J. Arnold Arbor. 27:423–437.

Cronquist A. 1968. The evolution and classification of flowering plants. London.

Cronquist A. 1988. The evolution and classification of flowering plants. New York.

Dandy J. E. 1964. Magnoliaceae. In J. Hutchinson, ed., The genera of flowering plants, pp. 50–57. Oxford.

Denffer D. von, 1983. Morphologie und Histologie des Cormus. In E. Strasburger et al., Lehrbuch der Botanik, 33d imp., 137–214. Stuttgart.

Deroin T. 1985. Contribution à la morphologie comparée du gynécée Annonaceae—Monodoroideae. Adansonia 2:167–176.

Dickson J. M. 1936. Studies in floral anatomy. III. An interpretation of the gynoecium in the Primulaceae. Amer. J. Bot. 33:385–393.

Douglas G. E. 1936. Studies in the vascular anatomy of the Primulaceae. Amer. J. Bot. 23:199–212.

Douglas G. E. 1944, 1957. The inferior ovary. Bot. Rev. 10:125–186; 23:1–46.

Eames A. J. 1931. The vascular anatomy of the flower with refutation of the theory of carpel polymorphism. Amer. J. Bot. 18:147–188.

Eames A. J. 1951. Again: "The New Morphology." New Phytol. 50:17–35.

Eames A. J. 1961. Morphology of the angiosperms. New York.

Eckardt T. 1937. Untersuchungen über Morphologie, Entwicklungsgeschichte und systematische Bedeutung des pseudomonomeren Gynoecium. Nova Acta Leop. Carol. (n.s.) 5:1–112.

Eckardt T. 1938. Das pseudomonomeren Gynoecium. Chron. Bot. 4:206–208.

Eckert G. 1966. Entwicklungsgeschichtliche und blütenanatomische Untersuchungen zum Problem der Obdiplostemonie. Bot. Jahrb. 85(4):523–604.

Ehrendorfer F. 1983. Samenpflanzen. In E. Strasburger et al., Lehrbuch der Botanik, pp. 758–915. Stuttgart.

Eichler A. W. 1875, 1878. Blütendiagramme. vols. 1, 2. Leipzig.

Endress P. K. 1982. Syncarpy and alternative modes of escaping disadvantages of apocarpy in primitive angiosperms. Taxon 31(1):48–52.

Endress P. K. 1987. Floral phyllotaxis and floral evolution. Bot. Jahrb. Syst. 108(2–3):417–438.

Endress P. I. 1990. Evolution of reproductive structures and functions in primitive angiosperms (Magnoliidae). Mem. New York Bot. Garden 55:5–34.

Endress P. K. and L. D. Hufford. 1989. The diversity of stamen structure and dehiscence patterns among Magnoliidae. Bot. J. Linn. Soc. 100:45–85.

Erbar C. 1983. Zum Karpellbau einiger Magnoliiden. Bot. Jahrb. Syst. 104(1):3–31.

Erbar C. 1986. Untersuchungen zur Entwicklung der spiraliger Blüte von *Stewartia pseudocamellia* (Theaceae). Bot. Jahrb. Syst. 106(3):391–407.

Erbar C. 1988. Early developmental patterns in flowers and their value for systematics. In P. Leins, S. C. Tucker, and P. K. Endress, eds., Aspects of floral development, pp. 7–23. Berlin and Stuttgart.

Erbar C. and P. Leins. 1981. Zur Spirale in Magnolien-blüten. Beitr. Biol. Pfl. 56:225–241.

Erbar C. and P. Leins. 1983. Zur Sequenz von blütenorganen bei einigen Magnoliiden. Bot. Jahrb. Syst. 103(4):433–449.

Eyde H. 1975. The bases of angiosperm phylogeny: floral anatomy. Ann. Missouri Bot. Gard. 62(3):521–537.

Foster A. S. and Gifford E. M. 1974. Comparative morphology of vascular plants. 2d ed. San Francisco.

Gelius L. 1967. Studien zur Entwicklungsgeschichte an Blüten der Saxifragales sensu lato mit besonderer Berücksichtigung des Androeceum. Bot. Jahrb. Syst. 87:253–303.

Goebel K. 1928. Morphologische und biologische Studien. XII-XV. Ann. Jard. Bot. Buitenzorg 39:1–232.

Goebel K. 1933. Organographie der Pflanzen. Part III. Samenpflanzen, 3d ed. Jena.

Grant V. 1950. The protection of the ovules in flowering plants. Evolution 4:179–201.

Grisebach A. 1854. Grundriss der Systematischen Botanik. Göttingen.

Gundersen A. 1950. Families of dicotyledons. Waltham, Massachusetts.

Hallier H. 1901. Uber die Verwandtschaftsverhältnisse der Tubifloren und Ebenalen, den polyphyletischen Ursprung der Sympetalen und Apetalen und die Anordnung der Angiospermen Uberhaupt. Vorstudien zum Entwurf eines Stammbaums der Blütenpflanzen. Abh. Naturwiss. Naturwiss. Verein Hamburg. 16(2):1–112.

Hallier H. 1902. Beiträge zur Morphogenie der Sporophylle und des Trophophylle in Beziehung zur Phylogenie der Kormophyten. Jahrb. Hamburg. Wiss. Anst. XIX, 3. Beiheft, 1–110.

Hallier H. 1903a. Vorläufiger Entwurf des naturlichen (phylogenetischen) Systems der Blütenpflanzen. Bull. Herb. Boissier II, 3:306–317.

Hallier H. 1903b. Uber die Abgrenzung und Verwandschaft der einzelnen Sippen bei den Scrophularineeen. Bull. Herb. Boissier II, 3(2): 181–207.

Hallier H. 1912. L'origine et le système phylètique des Angiosperms exposés à l'aide de leur arbre généalogique. Arch. Néerl. Sci. Exact. Nat. 3d ser., 1:146–234.

Hanf M. 1935. Vergleichend und entwicklungsgeschichte Untersuchungen über Morphologie und Anatomie der Griffel und Griffeläste. Beih. Bot. Zentralbl. 54A:99–141.

Harper J. L. and J. Ogden. 1970. The reproductive strategy of higher plants. I. The concept of strategy with special reference to *Senecio vulgaris* L. J. Ecol. 58(3):681–698.

Henslow G. 1891. On the vasuclar system of floral organs and their importance in the interpretation of the morphology of flowers. J. Linn. Soc. Bot. London 28:151–197.

Hiepko P. 1965a. Das zentrifugale Androceum der Paeoniaceae. Ber. Deutsch. Bot. Ges. 77:427–435.

Heipko P. 1965b. Vergleichend—morphologische und entwicklungsgeschichtliche Untersuchungen über das Perianth bei den Polycarpicae. I und II. Teil. Bot. Jahrb. 84:359–508.

Hofmeister W. 1868. Allgemeine Morphologie der Gevächse. Leipzig.

Howard R. A. 1948. The morphology and systematics of the West Indian Magnoliaceae. Bull. Torrey Bot. Club. 75(4):335–357.

Jackson G. E. 1934. The morphology of the flowers of *Rosa* and certain closely related genera. Amer. J. Bot. 21:453–466.

Joshi A. C. 1947. Floral histogenesis and carpel morphology. J. Indian Bot. Soc. 26:63–74.

Just Th. 1950. Carpels and ovules. In A. Gundersen, Families of dicotyledons, pp. 12–17. Waltham, Mass.

Kania W. 1973. Entwicklungsgeschichte Untersuchungen an Rosaceenblüten. Bot. Jahrb. Syst. 93:175–247.

Kozo-Poljanski B. M. 1922. An introduction to the phylogenetic systematics of the higher plants. Voronezh. (In Russian.)

Kozo-Poljanski B. M. 1923. On the systematical position of the family Compositae. J. Russian Bot. Soc. 8: 167–191. (In Russian.)

Kozo-Poljanski B. M. 1928. The ancestors of the angiosperms. Moscow. (In Russian.)

Laubengayer R. A. 1937. Studies in the anatomy and morphology of the polygonaceous flowers. Amer. J. Bot. 24:329–343.

Leins P. 1964a. Die frühe Blütenentwicklung von *Hypericum hookerianum* Wight et Arn. und *H. aegyptiacum* L. Ber. Deutsch. Bot. Ges. 77:112–123.

Leins P. 1964b. Das zentripetale und zentrifugale Androeceum. Ber. Deutsch. Bot. Ges. 77(71. Gen.-Vers):22–26.

Leins P. 1971. Das Androceum der Dikotylen. Ber. Deutsch. Bot. Ges. 84:191–193.

Leins P. 1972. Das Karpell im ober- und underständigen Gynoeceum. Ber. Deutsch. Bot. Ges. 85(7–9):291–294.

Leins P. 1975. Die Beziehungen zwischen multistaminaten und einfuchen Androeceen. Bot. Jahrb. Syst. 96:231–237.

Leins P. and C. Orth. 1979. Zur Entwicklungsgeschichte mänlicher Blüten von *Humulus lupulus* (Cannabaceae). Bot. Jahrb. Syst. 100:372–378.

Leroy J.-F. 1977. A compound ovary with open carpels in Winteraceae (Magnoliales). Evolutionary implications. Science 196:977–978.

Leroy J.-F. 1978. Une sous-famille monotypic de Winteraceae endémique a Madagascar: Les Takhtajanioideae. Adansonia. 2d ser., 17:383–395.

Lister G. 1883. On the origin of the placentas in the tribe Alisneae in the order Caryophylleae. Linn. Soc. J. Bot. 20:423–429.

Mayr B. 1969. Ontogenetische Studien an Myrtales-Blüten. Bot. Jahrb. 89:210–271.

Melchior H., ed. 1964. A. Engler's Syllabus der Pflanzenfamilien. 12th ed. vol. 2. Angiospermen. Berlin.

Merezhkovsky K. S. 1910. Conspective textbook of general botany. Kazan. (In Russian.)

Moseley M. F. 1958. Morphological studies in the Nymphaeaceae. I. The nature of stamens. Phytomorphology 8:1–29.

Nägeli C. von. 1884. Mechanisch-physiologische Theorie der Abstammungslehre. München und Leipzig.

Ozenda P. 1949. Recherches sur le Dicotylédones apocarpique. Contribution à l'étude des Angiospermes dites primitives. Paris.

Ozenda P. 1952. Remarques sur quelques interprétations de l'étamine. Phytomorphology 2:225–231.

Parkin J. 1914. The evolution of the inflorescence. J. Linn. Soc. Bot. London 42:511–553.

Parkin J. 1923. The strobilus theory of angiospermous descent. Proc. Linn. Soc. London 153:51–64.

Parkin J. 1951. The protrusion of the connective beyond the anther and its bearing on the evolution of the stamen. Phytomorphology 1:1–8.

Parkin J. 1955. A plea for a simpler gynoecium. Phytomorphology 5(1):46–57.

Payer J. B. 1857. Traité d'organogénie comparée de la fleur. Paris.

Philipson W. R. 1953. The relationships of the Compositae particularly as illustrated by the morphology of the inflorescence in the Rubiales and Campanulatae. Phytomorphology 3(4):391–404.

Philipson W. R. 1975. Evolutionary lines within the dicotyledons. New Zealand J. Bot. 13:73–91.

Pilger R.K. F. 1922. Über die Verzweigung und Blütenstandsbildung bei den Holzgewächsen. Bibl. Bot. 23:1–38.

Potonié H. 1912. Grundlinien der Pflanzenmorphologie im Lichte der Paläontologie. Jena.

Puri V. 1951. The role of floral anatomy in the solution of morphological problems. Bot. Rev. 17:471–533.

Puri V. 1952. Placentation in angiosperms. Bot. Rev. 18:603–651.

Rickett H. W. 1944. The classification of inflorescences. Bot. Rev. 10: 187–231.

Rickett H. W. 1955. Materials for a dictionary of botanical terms. III. Inflorescens. Bull. Torr. Bot. Club. 82:419–445.

Rohweder O. and P. K. Endress. 1983. Samenpflanzen. Morphologie und Systematik der Angiospermen und Gymnospermen. Stuttgart. New York.

Ronse Decraene L.-P. and E. Smets. 1987. The distribution and systematic relevance of the androecial characters oligomery and polymery in the Magnoliophytina. Nord. J. Bot. 7:239–253.

Sattler R. 1972. Centrifugal primordia inception in floral development. In Y. S. Murty et al., eds., Advances in plant morphology, pp. 170–178. (Prof. V. Puri commemoration vol.) Meerut.

Sattler R. 1973. Organogenesis in flowers. A photographic text-atlas. Toronto.

Sattler R. 1974. A new approach to gynoecial morphology. Phytomorphology 24(1, 2):22–34.

Sattler R. and F. Pauzé. 1978. L'androcée centripète *d'Ochna atropurpurea*. Canadian J. Bot. 56:2500–2511.

Sattler R. and V. Singh. 1977. Floral organogenesis of *Limnocharis flava*. Canadian J. Bot. 55:1076–1086.

Saunders E. R. 1936. Some morphological problems presented by the flowers in Nymphaeaceae. J. Bot. (London) 74:217–221.

Schaeppi H. and F. Steindl. 1942. Vergleichend - morphologische Untersuchungen an Gynoeceum der Rosoideen. Ber. Schweiz. Bot. Ges. 60:15–50.

Schaeppi H. and F. Steindl. 1945. Blütenmorphologische und

embryologische Untersuchungen an einigen Viscoideen. Vierteljahrsschr. Natuof. Ges. Zürich 90, Beih. 1:1–46.

Singh V. and R. Sattler. 1974. Floral development in *Butomus umbellatus*. Canadian J. Bot. 52:223–230.

Sinha S. C. and B. C. Joshi. 1959. Vascular anatomy of the flower of *Punica granatum*. Indian Bot. Soc. 38: 35–45.

Small J. 1919. The origin and development of the Compositae. Reprinted from New Phytol. vols. 16–18, 1917–1919.

Smith G. H. 1926, 1928. Vascular anatomy of ranalian flowers. Bot. Gaz. 82:1–29, 85: 152–177.

Smith F. H. and E. C. Smith. 1942. Floral anatomy of the Santalaceae and some related forms. Oregon State Monogr., Stud. Bot. 5:1–93.

Sprague T. A. 1925. The classification of dicotyledons. J. Bot., Brit. and Foreign 63:9–13, 105–113.

Stebbins G. L. 1974. Flowering plants. Evolution above the species level. London.

Stroebl F. 1925. Die Obdiplostemonie in den Blüten. Bot. Arch. 9:210–220.

Swamy B. G. L. 1949. Further contributions to the morphology of the Dengeneriaceae. J. Arnold. Arbor. 30:9–38.

Swamy B. G. L. and I. W. Bailey. 1950. *Sarcandra,* a vesselless genus of the Chloranthaceae. J. Arnold. Arbor. 31:117–129.

Takhtajan A. 1942. The structural types of gynoecium and placentation. Bull. Armen. Branch Acad. Sci. USSR 3–4 (17–18):91–112. (In Russian with English summary.)

Takhtajan A. 1948. Morphological evolution of the angiosperms. Moscow. (In Russian.)

Takhtajan A. 1959. Die Evolution der Angiospermen. Jena.

Takhtajan A. 1964. Foundations of the evolutionary morphology of angiosperms. Moscow and Leningrad. (In Russian.)

Takhtajan A. 1969. Flowering plants. Origin and dispersal. Edinburgh.

Takhtajan A. 1976. Neoteny and the origin of flowering plants. In C. B. Beck, ed., Origin and early evolution of angiosperms, pp. 207–219. New York.

Takhtajan A. 1980. Outline of the classification of flowering plants (Magnoliophyta). Bot. Rev. 46(3):225–359.

Takhtajan A. 1983. A revision of *Daiswa* (Trilliaceae). Brittonia 35:255–270.

Tamura M. 1963, 1965. Morphology, ecology, and phylogeny of the Ranunculaceae. I. Sci. Reports, Osaka Univ. 11:115–126, 14:53–71.

Tepfer S. S. 1953. Floral anatomy and ontogeny of *Aquilegia formosa* var. *truncata* and *Ranunculus repens*. Univ. Calif. Publ. Bot. 25:513–648.

Thompson B. F. 1942. The floral morphology of the Caryophyllaceae. Amer. J. Bot. 29:333–349.

Tiagi Y. D. 1955. Studies in floral morphology. II. Vascular anatomy of the flower in certain species of the Cactaceae. J. Indian Bot. Soc. 3:408–428.

Troll W. 1927. Zur Frage nach der Herkunft der Blumenblütter. Flora 122:57–75.

Troll W. 1928. Organisation und Gestalt im Bereich der Blüte. Berlin.

Troll W. 1939. Die morphologische Natur der Karpelle. Chron. Bot. 5:38–41

Troll W. 1954, 1957. Praktische Einführung in die Pflanzenmorphologie. I-II. Jena.

Troll W. 1964, 1969. Die Infloreszenzen. Typoligie und Stellung im Aufbau des Vegetationskörpers. Stuttgart.

Tucker S. C. 1984. Origin of symmetry in flowers. In: R. A. White and W. C. Dickinson, eds., Contemporary problems in plant anatomy, pp. 351–394. New York.

Van Tieghem Ph. 1868. Recherches sur la structure du pistil. Ann. Sci. Nat. 12:127–226.

Van Tieghem Ph. 1875. Recherches sur la structure du pistil et sur l'anatomie comparée de la fleur. Mém. Acad. Sci. Inst. Imp. de France 21:1–261.

Vink W. 1970, 1977. The Winteraceae of the Old World. I. *Pseudowintera* and *Drimys*—morphology and taxonomy. II. *Zygogynum*—morphology and taxonomy. Blumea 18:225–354, 23:219–250.

Vink W. 1978. The Winteraceae of the Old World: 3. Notes on the ovary Takhtajania. Blumea 24:521–525.

Weberling F. 1965. Typology of inflorescences. J. Linn. Soc. Bot. 59:215–221.

Weberling F. 1983a. Evolutionstendenzen bei Blütenständen. Abh. Akad. Wiss. Lit. Mainz, Math.-Naturw. Kl. Jg. 1983, Nr. 1.

Weberling F. 1983b. Fundamental features of modern inflorescence morphology. Bothalia 14(3,4):917–922.

Weberling F. 1988. Inflorescence structure in primitive angiosperms. Taxon 37:657–690.

Weberling F. 1989. Morphology of flowers and inflorescences. Translated by R. J. Pankhurst. Cambridge.

Wernham H. F. 1913 (1912). Floral evolution with particular reference to the sympetalous dicotyledons. New Phytol. Reprint 5:1–151.

Willemstein S. C. 1987. An evolutionary basis for pollination biology. Leiden.

Wilson C. L. and Th. Just. 1939. The morphology of the flower. Bot. Rev. 5:97–131.

Worsdell W. C. 1903. The origin of the perianth of flowers. New Phytol. 2:43–48

Zamyatin B. N. 1951. Bases of morphological interpretation of the inferior ovary. Ph.D. dissertation. Leningrad.

Zazhurilo K. K. and E. K. Kuznetsova. 1939. The nature of diffuse placentation. Trudy Voronezh. State Univ. 10(5):79–88. (In Russian.)

Zimmermann W. 1935. Die Phylogenie der Angiospermen—Blütenstände. Beih. Bot. Centralbl. 53(Abt. A):94–121.

Zimmermann W. 1965. Die Blütenstände, ihr System und ihre Phylogenie. Ber. Deutsch. Bot. Ges. 78:3–12.

3

Microsporangia, Microspores, and Pollen Grains

3.1. *The Microsporangium*

Stamens most commonly contain four microsporangia arranged in two pairs. Only in some taxa, such as Circaeasteraceae, Epacridaceae, certain Diapensiaceae, Bombacaceae, Malvaceae, Adoxaceae, Philydraceae, Restionaceae, the stamens contain only two microsporangia. Very rarely, as in *Arceuthobium* (Viscaceae) there is only one microsporangium. Multisporangiate stamens of some taxa, e.g., in Rhizophoraceae, result from partition of the sporogenous tissue by sterile plates.

The periclinal divisions of the hypodermal initials form a subepidermal layer of the subepidermal primary parietal cells and an inner layer of primary sporogenous cells. The cells of the primary parietal layer undergo further periclinal and anticlinal divisions leading to a variable number of concentrically arranged layers, which give rise to the mature microsporangium wall. The outermost layer next to the epidermis forms the so-called endothecium or "fibrous layer," the cells of which commonly bear fibrous thickenings. Below the endothecium, there is usually a layer (or several layers) of tubular thin-walled cells which represent the middle layers of the sporangial wall. The innermost layer of the sporangial wall surrounding the sporogenous tissue is the tapetum.

The endothecium, which is usually a single layer, reaches its full development at the time that the microsporangia are ready for dehiscence. The endothecium is not always present, especially in poricidal anthers, where the fibrous layer is usually absent, except in the region of the pores. It is absent also in certain members of Hydrocharitaceae and in some cleistogamous flowers.

There are usually one or two—rarely several—middle layers. The middle layers are ephemeral and—as a rule—become flattened, crushed, or obliterated at the time of the meiotic divisions. Very rarely, they are absent, as in *Vallisneria* and *Wolffia*.

The tapetum usually comprises a single layer, but sometimes it may divide and become biseriate and even multiseriate. The tapetal cells are full of dense cytoplasm. At the beginning of meiosis the cells of the tapetum enlarge and, in many cases, their nuclei may undergo some division. The food material entering into the sporogenous tissue passes through the tapetum and thus provides enzymes, hormones, and nutritive materials for the developing microsporocytes and microspores including the formation of exine and deposition of tryphine, a sticky golb of "pollenkitt" (pollen cement) and orbicules.

There are two structural and functional types of tapeta, distinguished on the basis of cell behavior during microsporogenesis: the secretory or glandular tapetum, the cells of which remain intact and persist in situ but, after meiosis, become disorganized and obliterated, and the periplasmodial or amoeboid tapetum, characterized by the breakdown of the cell walls before meiosis and protrusion of the protoplasts into the locule and fusion to form a multinucleate plasmodium. The overwhelming majority of families of flowering plants, including the majority of the most archaic taxa, is characterized by the secretory tapetum. In additions, some primitive characters are correlated with a secretory tapetum (Sporne 1973; Pacini et al. 1985). On the other hand, the periplasmodial type usually occurs in relatively more advanced groups. As Schürhoff (1926) pointed out, the presence of periplasmodial tapetum is closely correlated with an advanced character such as tricelled pollen grains.

The ways of dehiscence of the mature anther has also some systematic and evolutionary significance. The commonest and the most primitive dehiscence is the longitudinal dehiscence along the fissure (stomium), situated between a pair of microsporangia. The longitudinal dehiscence is of two types: by one simple longitudinal slit or by two longitudinal valves. The second type is characterized by additional, transverse slits usually at both ends of the longitudinal slit, which results in two windowlike lateral valves (see Endress and Hufford 1989 and Hufford and Endress 1989). Whereas the dehiscence by simple longitudinal slit is very common, the second type is characteristic of many Magnoliidae and Hamamelididae with more or less massive anthers and evidently derived from the first type. "Possibly, only the predisposition for easily developing valvate dehiscence was present in the original angiosperm stamen that dehisced via simple longitudinal slits. This predisposition would have been lost in more advanced angiosperms" (Endress and Hufford 1989:79). More specialized is a valvate dehiscence in Laurales and Berberidaceae, which typically arises by the opening of the thecal wall outward producing apically hinged flaps that lift upward at dehiscence. One of the most advanced types of dehiscence is the poricidal dehiscence, when pollen is released from a small opening situated at one end (distal or proximal). Examples of the latter are: Ochnaceae, Ericaceae, Myrsinaceae, some Fabaceae, the majority of Melastomataceae, Tremandraceae, Solanaceae. There are also other specialized modes of dehiscence including transverse dehiscence (e.g., *Alchemilla, Hibiscus, Euphorbia, Chrysosplenium*).

3.2. Microsporogenesis and Microspores

By dividing in various planes, the primary sporogeneous cells give rise to the microsporocytes, or microspore mother cells. In the early meiotic prophase, the original walls of the microsporocytes disintengrate, round off, and then separate from each other. A massive callose layer of polymer <<-1, 3 glucan is formed surrounding the microsporscytes. The microsporocytes undergo meiosis to produce the microspore tetrads.

The microspore tetrads are formed by two patterns determined by the mechanism of cytokinesis in microspore mother cells. In the successive type, the developing cell plate is formed at the end of meiosis I, dividing the microsporocyte into two cells; in each of these two cells, the second meiotic division takes place, followed again by centrifugal formation of cell plates. In the simultaneous type, on the other hand, no wall is formed after meiosis I; division occurs by centripetally advancing constriction furrows, which usually first appear after the second meiotic division, meet in the center, and divide the mother cell into four parts. The constriction furrows originate at the surface of the mother cell and develop inwardly, resulting in the formation of walls that divide the microsporocyte into four microspores.

It is difficult to say which of the two types of microsporogenesis is more primitive. Although some authors (including Schürhoff 1926 and Davis 1966) consider the successive type as the more primitive, there is no definite correlation between this type and archaic Magnoliidae and Ranunculidae. The majority of Magnoliidae and Ranunculidae are characterized by simultaneous microsporogenesis. The successive type is known in some Annonaceae (*Cananga*), in Myristicaceae, Monimiaceae, Lauraceae, Hernandiaceae, in some Aristolochiaceae, in *Cabomba*, in some species of *Ceratophyllum, Mahonia,* and *Thalictzum* (Yakovlev 1981), but all other members of these archaic subclasses are characterized by the simultaneous type. Successive microsporogenesis is characteristic also for such advanced magnoliopsids as some Apocynaceae and the family Asclepiadaceae and many liliopsids including Butomaceae, Hydrocharitaceae, Alismatales, Najadanae, Triuridaceae, and the majority of Lilianae.

3.3. Pollen Grains

Prior to the dehiscence of the anther, each of haploid microspores gives rise to a two- or three-celled endosporic male gametophyte. Thus, the micropsores give rise to pollen grains, which are the carriers of the male gametophytes.

Usually, the pollen grains of each tetrad separate from each other quickly while still in the ripe anther and are shed as monads. But in some families (e.g., Winteraceae, Nepenthaceae, and Ericaceae), the pollen grains are shed while still joined in the tetrads. A special type of strongly modified tetrad is provided by the pseudomonads (cryptotetrads) of Cyperaceae and Epacridaceae-Styphelieae, where three nuclei of each tetrad degenerate, and only one micropsore develops. In addition to the tetrads, dyads, and polyads (multiples of tetrads) or more complex masses—pollinia—are found in some taxa. But in the vast majority of flowering plants, the microspores (pollen grains) are isolated from each other as monads.

The pollen grains of flowering plants are distinguished by their unusual diversity in size, shape, and structure. Though their sizes are very varied, for each species they are commonly constant enough and vary only within certain limits. As a general rule, the less advanced families more often have large pollen grains; whereas, in more specialized sympetaleous families, a trend towards the diminution in their size is noted. Although during evolution pollen grains commonly decreased in size, often the reverse process of increase is also observed. Thus in some Nyctaginaceae, Cucurbitaceae, Malvaceae, or in the genus *Morina,* the pollen grains are so large that they can be distinguished by the naked eye.

A certain relationship is observed between the pollen grain size and that of the flower. Large grains are found more often in plants with large flowers, whereas small grains dominate in the plants with small flowers aggregated in an inflorescence. With a decrease in the general size of the flower, naturally the size of the anthers also decreases, leading to a diminution in the size of the individual pollen grains. But a much more important factor determining the decrease of the pollen grain size is the diminution of the gynoecium and the reduction of the length of the stylodia or styles and, consequently, the shortening of that distance which the pollen tube should cover from the stigma to the female gametophyte. Large size of the pollen grains in species with long stylodia or styles is explained by the necessity of accumulation of a larger reserve of nutritive substances required for the growth of the pollen tubes. It should, however, be noted that the ratio of the pollen grain size and the distance covered by the pollen tube is fairly constant within a taxon, often within a family, but is variable among different families and even different genera. Presumably, there are other factors as well which determine the size of the pollen grain.

The pollen wall, as a rule, consists of two main layers—the inner one, called intine, and the outer one, called exine. These two layers are chemically, structurally, and developmentally distinct.

The intine is a hyaline extra-cytoplasmic membrane of microfibrillar structure which is synthesized by a microspore protoplast soon after the release of microspores from the tetrad (Heslop-Harrison 1968). It appears only after the exine is nearly mature. Intine growth involves the activity of both dictyosomes and endoplasmatic reticulum (Gabarayeva 1986b, 1987c). In certain archaic magnoliopsids—for instance, in *Michelia fuscata, Manglietia tenuipes, Magnolia delavayi*, and *Liriodendron tulipifera*—there is a compound three-layered intine, of which the first layer is granular and is formed exclusively by the Golgi apparatus, whereas the second and third layers are formed by both Golgi apparatus and endoplasmatic reticulum (Gabarayeva 1986b, 1987c). The intine forms a continuous coat around the protoplast and is devoid of apertures. It consists mainly of cellulose but it may contain proteins and glycoproteins stored in its polysaccharide matrix (Knox 1984). The thickness of intine varies greatly in different taxa: in some taxa this layer is too thin to be observed with the optical microscope, while in others it is very thick. It reaches its greatest development and complexity in the apertural regions of the exine, e.g., in *Malvaviscus* (Heslop-Harrison et al. 1973). In some taxa with very reduced exine, e.g., in many Zingiberales, the intine is most often highly elaborated, stratified, and chaneled (Kress and Stone 1982; Kress 1986).

The exine is a complex multifunctional structure. Its three main functions are: 1) the protection function—protection of the male gametophyte and prevention of its dessication during pollination (Wodehouse 1935; Payne 1972, 1981; Muller 1979; Kress 1986); 2) the reservation function—storage of recognition compounds involved in pollen germination and pollen-stigma interactions (Heslop-Harrison 1975, 1976, 1978; Heslop-Harrison et al. 1975; Knox et al. 1975); and 3) the clustering function, related to mode of dispersal (Muller 1979). The exine typically consists of two layers—the inner layer endexine and the outer layer ectexine. The two layers differ chemically, structurally, and developmentally. They are very distinct in the transmission electron microscope.

The endexine (nexine 2, according to Erdtman 1969; or secondary exine, according to Godwin et al. 1967) probably comprises both polysaccharides and sporopollenin. In gymnospermous plants, the endexine is generally laminated (Van Campo 1971) with a series of parallel laminations that are continuous throughout both the apertural and nonapertural regions of the endexine (Walker and Walker 1984). In flowering plants, the endexine may be found as a continuous layer (sometimes very thick, as in Lauraceae) or only in apertural regions. In some taxa, endexine is absent. Usually, the endexine is not laminated, but in some taxa as *Sorghum bicolor* (Christensen and Horner 1974), *Silene alba* (Shoup et al. 1981) and *Michelia fuscata, Manglietia tenuipes, Magnolia delavayi, Liriodendron chinensis, Anaxagorea brevipens, Asimina triloba* (Gabarayeva 1986a, 1986b, 1987a–c, 1988), the endexine is tangentially lamellated (especially under the apertures). The endexine frequently has wide pores and wide channels (as in Malvaceae) and also some fine sculpturing on its inner face (Van Campo 1971; Le Thomas and Lugardon 1972). In ontogeny, the endexine is produced later than the ectexine. Developmentally and genetically, the endexine is nearer to the intine than to the ectexine. It develops in young free microspores on the outer surface of its plasmalemma (Barnes and Blackmore 1986). As Gabarayeva (1986a, 1986b) has recently shown, the endexine lamellae in *Michelia fuscata* are formed in the course of budding of the vesicules from the microspore plasma membrane. It is interesting to note that in some taxa, the endexine is interbedded with the intine (Zavada 1984). It is not surprising, as both these two layers have common origin.

With the endexine, the so-called onci (sing. oncus) are closely connected. According to Praglowski and Raj (1979:110), an oncus is an intinuous thickening, biconvex or plano-convex in lateral view and circular when seen from above, underlying a pore or circular os. But according to Rowley and Dahl (1977) and Knox (1984), ultrathin sections show that the oncus is formed from endexine and is a complex lamellar structure. Later in pollen development, it regresses, and it is absent by the maturation period. The oncus may play a part in nourishing the pollen tube by dissolving when the pollen grain germinates (Faegri and Iversen 1989). According to Barnes and Blackmore (1986:78), the oncus may also function during germination as a preformed tip to the emerging pollen tube.

In an overwhelming majority of flowering plants the ectexine is well developed and stratified. It is an acetolysis-resistant layer formed of one of the most extraordinary chemically and biologically resistant materials known in the organic world—sporopollenin, which is the product of oxydative polymerization of carotenoids and carotenoid esters (Shaw 1971; Brooks and Shaw 1978). It is known to be free of nitrogen and contains no cellulose. It has many properties in common with lignin and cutin.

The exine structure and ornamentation (sculpturing) is extremely varied and, at the same time, very constant within the taxonomic groups and has a large systematic and evolutionary significance.

The ectexine consists of two basic layers—a rooflike outer layer or tectum and an infratectal layer. The latter is of two main types—granular and columellar. Granular structure occurs in a wide variety of both gymnosperm and angiosperm pollen grains, with an infratectal layer consisting of more or less densely aggregated, equidimensional granules of sporopollenin (Van Campo and Lugardon 1973; Doyle, Van Campo, and Lugardon 1975). The tectum, which is not always noticeable, is composed of more densely aggregated granules. Doyle et al. (1975:436) suspect that at least some of the apparently homogeneous "atectate" exine of Walker and Skvarla, revealed in some of the most archaic Magnoliidae such as *Degeneria* and *Eupomatia*, are extreme members of the granular category, with very closely aggregated granules. The infratectal layer in some Magnoliidae is most probably primitively granular. But the granular structure is also found in many advanced groups such as *Casuarina, Alnus, Betula, Carpinus, Myrica, Rhoiptelea, Juglans, Engelhardtia, Carya*, Batales, and various members of Rosidae and Asteridae. I agree with Doyle et al. (1975:439), that the occurrence of granular structure in such putatively advanced groups "suggests that the granular-columellar trend is frequently reversed in higher dicots (and not just aquatics, parasites, etc., cited by Walker and Skvarla, 1975)."

The predominant type of infratectal structure is columellar. It is characterized by radially directed rods of lineary fused sporopollenin granules, the columellae, which are commonly supported by the homogeneous or laminated (as in certain Annonaceae) basal layer or sole (foot-layer). In great majority of flowering plants, including the majority of Magnoliidae, the heads of the columellae extend laterally over the intercolumellar spaces forming a rooflike layer, the tectum. Comparative studies of the ectexine ultrastructure suggest an evolutionary trend from granular ectexine to incipient rudimentary columellae and from the incipient columellae to fully developed columellar structure. I agree with Walker and Skvarla (1975) that the columellae evolved independently a number of times even within, for example, different subfamilies of the Annonaceae.

In the most primitive type of columellar ectexine (figure 18) the tectum is devoid of any kind of holes or perforations (Walker 1974a). This tectate-imperforate (Walker 1974a) or completely tectate ectexine (Hideux and Ferguson 1976) is found in various groups of flowering plants both archaic and advanced. The completely tectate ectexine has either a more or less even surface (psilate tectum) or is supra-ornate, that is, covered with various kinds of supratectal ornamentation (sculpturing): pits, grooves,

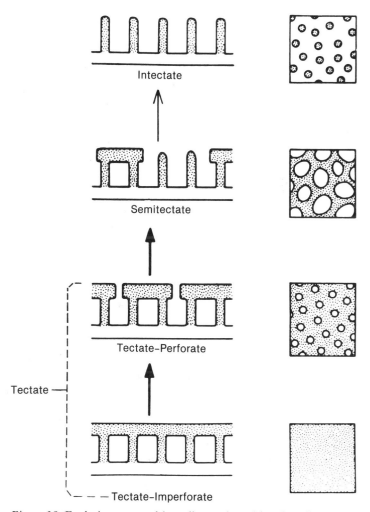

Figure 18. Evolutionary trend in pollen grains with columellate ectexine (from Walker 1974).

or radial projections such as granules, warts, spines, networks, combs, ridges, and so forth. The granular or granular-verrucate tectal surface is probably the most primitive type of supratectal ornamentation (Takhtajan 1948:204), like that of some Magnoliaceae, e.g., *Aromadendron.*

The next evolutionary stage of the tectum structure is the perforate type (Walker 1974a, Hideux and Ferguson 1976). In the perforate tectum, the holes or tectal perforations (lumina) are always small (e.g., in some Annonaceae and Myristicaceae) and the columellae are invisible through them. When perforations enlarge so that their diameter becomes greater than the width of the pollen wall between them (muri), e.g., in Winteraceae, Illiciaceae, and Schisandraceae, the exine becomes semitectate (Walker 1974a). For this partial tectum, the visibility of columellae in oblique view through the lumina is characteristic (Hideux and Ferguson 1976). In semitectate exine, a network of muri may connect the heads of the columellae and form an open reticulum (perreticulate exine).

When the tectum is completely lost, e.g., in some Annonaceae, Myristicaceae, and Salicaceae, and there are only free, exposed columellae or their modified derivatives, we have intectate exine (Walker 1974a). Surface elements present in intectate (and semitectate) pollen, such as the reticulum, the rodlike columellae etc., are defined by Walker (1974a) as comprising simultaneously, both structure and sculpturing (ornamentation). They are "columella-derived structures, mostly with a reticulate architecture" (Muller 1979:601).

The culmination of an evolutionary trend is the origin of the almost exineless pollen with a much expanded and highly structured intine. According to Kress (1986), pollen grains possessing a very reduced exine and much elaborated intine are known to occur in unrelated families of flowering plants scattered both in magnoliopsids and liliopsids. They are known even in some Magnoliidae, including some Annonaceae (Le Thomas et al. 1986) and Lauraceae (Hesse and Kubitzki 1983).

Most pollen grains have specially delimited apertures—generally thin-walled areas or openings in the exine which serve as exits through which the pollen tubes usually emerge. Apertures manifest themselves in both the ect- and endexine, although to a varying degree (Faegri and Iversen 1975). In some grains, the ect- and endexinous apertures are concordant, in others not.

The apertures of the flowering plants pollen grains are characterized by a great diversity and are of various types. The arrangement, shape, and structure of the apertures have an exceptional significance and are very often used for solving some problem or the other of the magnoliophyte systematics. Various types of apertures correspond to different levels of specialization, and the significance of these types is very important in determining the general level of organization of some taxon or other.

The surface pattern of the microspore is related to its orientation on the tetrad stage (Wodehouse 1935). Two opposite polar parts of each microspore are called respectively proximal and distal, the line between them is called polar axis, and the plane perpendicular to the axis is the equatorial plane. In their arrangement, the apertures may be distal (but never proximal, in monads), zonal (with the centers at the equator or on one of the lines parallel to the equator) and global (more or less uniformly distributed over the entire surface). The apertural arrangement in the angiosperm pollen grains evolved from distal through zonal to global.

As long ago as 1912, Hans Hallier concluded that the most primitive type of pollen grain is characterized "par une seul por germinal," by which he apparently meant furrow and not a pore in the strict sense of the word. Later, it was shown that the most primitive angiosperm pollen grain is a type with one distal germinal furrow (distal colpus or "sulcus") in the exine (Wodehouse 1936; Bailey and Nast 1943; Takhtajan 1948, 1959; Canright 1953; Eames 1961; Doyle 1969; Muller 1970, 1979; Agababian 1973; Stebbins 1974; Walker 1974b, 1976a, 1976b; Straka 1975; and many others).

The pollen grain with one distal colpus is the only type found in both the magnoliophytes and gymnospermous groups. It characterizes a number of fossil gymnosperms, while it is most clearly expressed in Cycadales and *Ginkgo* among the recent gymnosperms. On the other hand, almost all the representatives of the superorder Magnolianae have monocolpate pollen grains, which also characterize the order Nymphaeales, the families Saururaceae and Piperaceae, the archaic genus *Saruma* of Aristolochiaceae, Hydnoraceae, some members of Rafflesiales, and most of Liliopsids. The pollen grains of Magnoliaceae are still very similar to the pollen grains of Cycadeoideales, Cycadales, and *Ginkgo,* and the distinction here lies essentially in the details.

In some cases, as for example in the family Lauraceae, a reduction of the distal colpus took place, resulting in an inaperturate pollen grain. The intine here is much thickened and the exine is transformed into a tender transparent membrane and consequently the function of the colpus is executed by the entire pollen grain surface. At the time of germination, the pollen tube is able to emerge from any point on the surface of the grain. In terms of function, the entire sporoderm is an aperture (Kress 1986:331). The reduction of the distal colpus occurred also in a number of aquatic liliopsids where it was, however, realized not by the thickening of the intine but due to the thinning and finally the disappearance of the exine in connection with hydrophily. In the family Hydrocharitaceae, the pollen grains of the genus *Vallisneria* still have the reduced colpus, but in *Stratiotes, Elodea, Hydrocharis,* and *Ottelia* the colpus has already disappeared completely. The reduction of the exine and the disappearance of the colpus connected with it is very well expressed in Scheuchzeriaceae, Juncaginaceae, Posidoniaceae, Zosteraceae, *Trillium,* the majority of Zingiberales, and some others.

While in hydrophilous liliopsids the distal aperture tends to disappear, in many anemophilous liliopsids it is only shortened and more or less reduced, thus being transformed into a distal pore. Such monoporate pollen grains characterize, for example, the Flagellariaceae, Joinvilleaceae, Restionaceae, Centrolepida-ceae, Poaceae, and others. They attained the highest level of specialization in grasses.

In some groups, the distal colpus assumes the form of a three-armed (rarely four-armed and in species of *Hedyosmum* even six- to seven-armed) aperture. These trichotomocolpate pollen grains are found in some Magnoliidae such as Annon-aceae, Canellaceae, and Chloranthaceae (*Hedyosmum*), and in certain liliopsids such as some Tecophilaeaceae, Asphodelaceae, and Arecaceae.

In some groups of flowering plants, the pollen grains with a single distal colpus may evolve into those with a larger number of distal colpi. Thus, in the Atherospermataceae, Calycantha-ceae, *Idiospermum, Tofieldia, Rigidella, Tigridia,* certain Agava-ceae, Amaryllidaceae, Pontederiaceae, Dioscoreaceae, Araceae, and some others, the pollen grains have two distal colpi, whereas in some Monimiaceae, Liliaceae, Pontederiaceae, Dioscorea-ceae, Haemodoraceae, and Araceae, three colpi are also found, and, in certain cases—as in the Dioscoreaceae—even more. With the increases in the number of distal colpi, usually their disposition is already not restricted to the polar region but shifts more or less toward the equator.

Another transformation of the distal colpus is its shift to zonal position parallel to the equator as in some Nymphaeaceae (*Nymphaea, Ondinea, Euryale, Victoria*) or around the equator as in *Eupomatia*. This type of monocolpate pollen grains is usually called zonosulculate (Erdtman 1952 and others), but I would prefer the more appropriate term cyclocolpate.

In some cases, the typical monocolpate pollen grains trans-formed into the grains with spiral colpi (spiraperturate pollen grains). Spiraperturate pollen grains are found in some liliop-sids, such as *Lomandra, Aphyllanthes,* and Eriocaulaceae. A mod-ified type of the spiraperturate pollen grain is observed in *Bax-teria* (Dasypogonaceae).

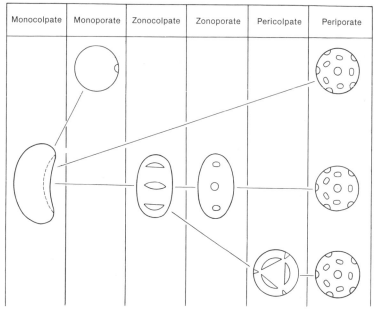

Figure 19. Diagram of the evolution of main types of apertures (from A. Takhtajan 1959).

The major evolutionary trend among monocolpate derived apertures is the origin of porate apertures (figure 19). In the family Winteraceae, the distal aperture is more or less porelike, ulcerate according to Erdtman's (1952) terminology. But monoporate (including "ulcerate") pollen grains are especially widespread among the liliopsids. They are characteristic for many Cyperaceae and for Restionales, Poales, Typhales, and others. More advanced are zonoporate pollen grains (several pores in an equatorial position) and periporate (panporate) pollen grains (usually a larger number of pores in a global position). There are no zonoporate pollen grains among the Magnoliidae. In the liliopsids, they occur only in *Conostylis setosa* (Conostylidaceae) and *Sclerosperma mannii* (Arecaceae). Periporate pollen grains are found in certain Trimeniaceae, in some species of *Aristolo-*

chia, and in many liliopsids, especially within the subclass Alismatidae.

The most important evolutionary change in pollen grains is the transition from primary monocolpate grains to zonocolpate ones with meridional (perpendicular to the equator) furrowlike apertures. This transition took place only among the magnoliopsids. The apertures of these primitive zonocolpate pollen grains, like those with distal or distally derived apertures are simple, characterized by continuous and more or less uniform aperture membranes. The most primitive type of zonocolpate pollen grains are tricolpate grains. Such primitive tricolpate pollen grains are found in the Illiciaceae, some Schisandraceae, *Nelumbo,* most of the Lardizabalaceae, *Sargentodoxa,* many Menispermaceae and Ranunculaceae, Circaeasteraceae, *Hydrastis,* many Berberidaceae, *Nandina, Glaucidium,* most of the Papaveraceae, certain Fumariaceae, many Caryophyllidae, *Trochodendron, Tetracentron, Cercidiphyllum,* and in many others. In many lines of evolution, tricolpate pollen grains gave rise to quadri-, hexa-, and polycolpate ones. They are found in very diverse families, but frequently also in one and the same family and often within the same genus. Thus in the family Papaveraceae some genera like *Arctomecon* and *Argemone* have tricolpate pollen grains, the genus *Dendromecon* is characterized by tri-pentacolpate pollen grains, while *Eschscholtzia* has from four to seven colpi and *Hunnemannia* has nine. A similar process of polymerization of the meridional colpi is also observed in some other families including Lamiaceae. This indicates that the transition of the tricolpate type into the polycolpate is not related to such a sharp change in the sporoderm structure as the transition of the polar type into the zonocolpate.

Another direction of evolution of the tricolpate pollen grains is provided by the reduction in the number of colpi to two. Such dicolpate pollen grains characterize, for example, *Hype-*

coum. In certain groups, as in some Brassicaceae, the inaperturate pollen grains evolved from the tricolpate ones.

In certain groups of magnoliopsids, zonocolpate pollen grains with colpi arranged along the meridians gave rise to pericolpate pollen grains. The colpi in the pericolpate pollen grains are oriented in different directions and evenly distributed over the whole surface of the grain. The pericolpate pollen grains are quite widespread and found in the representatives of the most diverse families. They are found in certain genera of the Ranunculaceae, in a number of Papaveraceae, Phytolaccaceae, Portulacaceae, Cactaceae, Caryophylaceae, *Euptelea*, and others. The number of colpi at times reaches thirty.

Just as monocolpate pollen grains gave rise to monoporate and polyporate forms, the tricolpate and polycolpate types gave rise to the triporate and polyporate pollen grains. They are found in certain Ranunculaceae (such as *Thalictrum, Anemone, Clematis, Ranunculus*), Papaveraceae (such as *Sanguinaria*), many Caryophyllales, *Altingia*, and others.

All the three pores in the triporate pollen grains are situated along the equator and resulted from the reduction of the three meridional colpi. Such pollen grains characterize, for example, many Urticales, some Campanulaceae, and many other magnoliopsids. Polyporate pollen grains are of two types. In one variant, as, e.g., in *Urtica*, the pores are situated along the equatorial line. Such pollen grains evolved from the zonocolpate forms. But in most cases the pores are distributed over the entire surface of the grain. Such periporate pollen grains characterize, for example, *Liquidambar, Buxus*, most Caryophyllaceae, Chenopodiaceae, and others. In a number of cases, the periporate pollen grains probably evolved from the zonoporate pollen grains by an increase in the number of pores.

In all the above-mentioned pollen grain types, the aperture is simple. But in the majority of angiosperm pollen grains, the aperture is compound. The compound or orate aperture possesses delimited areas of the aperture membrane known as ora (singular, os). There are as many types of orate pollen grains as nonorate ones—zonopororate (tri-polypororate) and peripororate. There is a noticeable parallelism in variation and evolution of nonorate and orate apertures. Orate apertures evolved from the nonorate ones.

Such are the main lines of the evolution of pollen grains as it appears to us at present.

References

Agababian V. Sh. 1973. Pollen grains of primitive angiosperms. Erevan. (In Russian.)

Bailey I. W. and C. G. Nast. 1943. The comparative morphology of the Winteraceae. I. Pollen and stamens. J. Arnold. Arbor. 24:340–346.

Barnes S. H. and S. Blackmore. 1986. Some functional features in pollen development. In S. Blackmore and I. K. Ferguson, eds., Pollen and function, pp. 71–80. (Linn. Soc. Symp. ser. 12.) London.

Blackmore S. and S. H. Barnes. 1986. Harmomegathic mechanisms in pollen grains. In S. Blackmore and I. K. Ferguson, eds., Pollen and spores. Forms and function, pp. 137–149. (Linn. Soc. Symp. ser. 12.) London.

Brooks J. and G. Shaw. 1978. Sporopollenin: a review of its chemistry, palaeochemistry, and geochemistry. Grana 17:91–97.

Canright J. E. 1953. The comparative morphology and relationships of the Magnoliaceae. II. Significance of the pollen. Phytomorphology 3:355–365.

Christensen J. E. and H. T. Horner, Jr. 1974. Pollen development and its spatial orientation during microsporogenesis in the grass *Sorghum biicolor*. Amer. J. Bot. 61:604–623.

Davis G. L. 1966. Systematic embryology of the angiosperms. New York, London, and Sydney.

Doyle J. A. 1969. Cretaceous angiosperm pollen of the Atlantic Coastal Plain and its evolutionary significance. J. Arnold Arbor. 50:1–35.

Doyle J. A., M. Van Campo, and B. Lugardon. 1975. Observations on exine structure of *Eucommidites* and Lower Cretaceous angiosperm Pollen. Pollen Spores 17:429–486.

Eames A. J. 1961. Morphology of the angiosperms. New York.

Endress P. K. and L. D. Hufford. 1989. The diversity of stamen structure and dehiscence patterns among Magnoliidae. Bot. J. Linn. Soc. 100:45–85.

Erdtman G. 1952. Pollen morphology and plant taxonomy: Angiosperms. Stockholm.

Erdtman G. 1969. Handbook of palynology. Copenhagen.

Esau K. 1977. Anatomy of seed plants. New York.

Faegri K. and J. Iversen. 1989. Textbook of pollen analysis. IV edition by K. Faegri, P. E. Kaland, and K. Krzywinski. Chichester.

Gabarayeva N. I. 1986a. The development of the exine in *Michelia fuscata* (Magnoliaceae) in connection with the change in cytoplasmic organelles of microspores and tapetum. Bot. Zhurn. (Leningrad) 71(3):311–322. (In Russian with English summary.)

Gabarayeva N. I. 1986b. Ultrastructure analysis of the intine development of *Michelia fuscata* (Magnoliaceae) in connection with the changes of cytoplasmic organelles of microspores and tapetum. Bot. Zhurn. (Leningrad) 71(4):416–428. (In Russian with English summary.)

Gabarayeva N. I. 1987a. Ultrastructure and development of sporoderm in *Manglietia tenuipes* (Magnoliaceae) during tetrad period: the primexine formation in connection with cytoplasmic organelle activity. Bot. Zhurn. (Leningrad) 72(3):281–290. (In Russian with English summary.)

Gabarayeva N. I. 1987b. Ultrastructure and development of lamellae of endexine in *Manglietia tenuipes* (Magnoliaceae) in connection with the question of endexine existence in primitive angiosperms. Bot. Zhurn. (Leningrad) 72(10):1310–1317. (In Russian with English summary.)

Gabarayeva N. I. 1987c. Ultrastructure and development of pollen grain wall in *Manglietia tenuipes* (Magnoliaceae): the formation of intine in connection with the activity of cytoplasmic organelles. Bot. Zhurn. (Leningrad) 72(11):1470–1478. (In Russian with English summary.)

Gabarayeva N. I. 1988. The significance of ontogenetic investigation of sporoderm for elucidation of structure and phylogeny of the mature sporoderm of some species of Magnoliaceae and Annonaceae. In A. F. Chlonova, ed., Palynology in the USSR. Novosibirsk, pp. 48–52. (In Russian with English summaries).

Godwin H. 1968. The origin of the exine. New Phytol. 67:667–676.

Godwin H., P. Echlin, and B. Chapman. 1967. The development of the pollen grain wall in *Ipomoea pupurea* (L.). Roth. Rev. Palaeobot. Palynol. 3:181–195.

Hallier H. 1912. L'origine et la système phylétique des Angiosperms exposés à l'aide de leur arbre généalogique. Arch. Néerl. Sci. Exact. Nat. ser. 3 B, 1:146–234.

Heslop-Harrison J. 1968. The pollen grain wall. Science 161:230–237.

Heslop-Harrison J. 1975. Incompatibility and the pollen-stigma interaction. Rev. Plant Physiol. 26:403–425.

Heslop-Harrison J. 1976. Adaptive significance of the exine. In I. K. Ferguson and J. Muller, eds., The evolutionary significance of the exine, pp. 25–37. Linn. Soc. Symp. ser. 1. London.

Heslop-Harrison J. 1978. Cellular recognition systems in plants. Studies in biology, no. 100. London.

Heslop-Harrison J., Y. Heslop-Harrison, J. and Barber. 1975. The stigma surface in incompatibility responses. Proc. Roy. Soc. B, 188:287–297.

Heslop-Harrison J., Y. Heslop-Harrison, R. B. Knox, and B. J. Howlett. 1973. Pollen-wall "gametophytic" and "sporophytic" fractions in the pollen walls of the Malvaceae. Ann. Bot. 37:403–412.

Hesse M. and K. Kubitski. 1983. The sporoderm ultrastructure in *Persea, Nectandra, Hernandia, Gomortega,* and some other lauralean genera. Plant Syst. Evol. 141:299–311.

Hideux M. J. and I. K. Ferguson. 1976. The stereostructure of the exine and its evolutionary significance in Saxifragaceae sensu lato. In I.K. Ferguson and J. Muller, eds., The evolutionary significance of the exine, pp. 327–377. Linn. Soc. Symp. ser. 1. London.

Hufford L. D. and P. K. Endress. 1989. The diversity of anther structures and dehiscence patterns among Hamamelididae. Bot. J. Linn. Soc. 99:301–346.

Knox R. B. 1984. The pollen grain. In B. M. Johri, ed., Embryology of angiosperms, pp. 197–271. Berlin.

Knox R. B., J. Heslop-Harrison, and Y. Heslop-Harrison. 1975. Pollen-wall proteins: localisation and characterization of gametophytic and sporophytic fractions. Biol. J. Linn. Soc. 7:177–187.

Knox R. B. and C. A. McConchie. 1986. Structure and function of compound pollen. In S. Blackmore and I. K. Ferguson, eds., Pollen and spores. Form and function. pp. 265–282. Linn. Soc. Symp. ser. 12. London.

Kress W. J. 1986. Exineless pollen structure and pollination systems of tropical *Heliconia* (Heliconiaceae). In S. Blackmore and I. K. Ferguson, eds., Pollen and spores. Form and function, pp. 329–345. Linn. Soc. Symp. ser. 12. London.

Kress W. J. and D. E. Stone. 1982. Nature of the sporoderm in monocotyledons, with special reference to the pollen grains of *Canna* and *Heliconia*. Grana 21:129–148.

Le Thomas A. 1980, 1981. Ultrastructural characters of the pollen grains of African Annonaceae and their significance for the phylogeny of primitive Angiosperms. I-II. Pollen Spores 22:267–342, 23:5–36.

Le Thomas A. and B. Lugardon. 1972. Sur la structure fine des tétrades de deux Annonacées (*Asteranthe asterias* et *Hexalobus monopetalus*). Compte Rendus Séances Acad. Sci., Paris (ser. D) 275:1749–1752.

Le Thomas A. and B. Lugardon. 1976. De la structure grenue à la structure columellaire dans le pollen des Annonacées. Adansonia ser. 2, 15:543–572.

Le Thomas A., W. Morawetz, and Waha M. 1986. Pollen of palaeo- and neotropical Annonaceae: definition of the aperture by morphological and functional characters. In S. Blackmore and I. K. Ferguson, Pollen and spores. Form and function, pp. 375–388. Linn. Soc. Symp. ser. 12. London.

Le Thomas A. 1988. Variation de la région aperturale dans le pollen des Annonacées. Taxon 37(3):644–656.

Muller J. 1970. Palynological evidence on early differentiation of angiosperms. Biol. Rev. 45:417–450.

Muller J. 1979 (1980). Form and function in angiosperm pollen. Ann. Missouri Bot. Gard. 66:593–632.

Nowicke J. W. and J. J. Skvarla. 1979 (1980). Pollen morphology: the potential influence in higher order systematics. Ann. Missouri Bot. Gard. 66:633–700.

Pacini E., G. G. Franchi, and M. Hesse. 1985. The tapetum: its form, function, and possible phylogeny in Embryophyta. Plant Syst. Evol. 149:155–185.

Payne W. W. 1972. Observations of harmomegathy in pollen of Anthophyta. Grana 12:93–98.

Payne W. W. 1981. Structure and function in angiosperm pollen wall evolution. Rev. Palaeobot. Palynol. 35:39–50.

Praglowski J. 1974. Magnoliaceae Juss. World Pollen and Spore Flora. 3:1–45.

Praglowski J. and B. Raj. 1979. On some pollen morphological concepts. Grana 18:109–113.

Rowley J. R. 1978. The origin, ontogeny and evolution of the exine. IV. Intern. Palynol. Conf., Lucknow (1976–77) 1:126–136.

Rowley J. R. and A. O. Dahl. 1977. Pollen development in *Artemisia vulgaris* with special reference to glycocaly material. Pollen Spores 19:169–284.

Schürhoff P. N. 1926. Die Zytologie der Blütenpflanzen. Stuttgart.

Shaw G. 1971. The chemistry of sporopollenin. In J. Brooks, P. R. Grant, M. D. Muir, G. Shaw, and P. van Gijzel, eds., Sporopollenin, pp. 305–348. London.

Shoup J.R., J. Overton, and M. Ruddat. 1981. Ultrastructure and development of the sexine in the pollen wall of *Silene alba* (Caryophyllaceae). Bot. Gaz. 141:379–388.

Sporne K. R. 1973. A note on the evolutionary status of tapetal types in dicotyledons. New Phytol. 72:1173–1174.

Stebbins G. L. 1974. Flowering plants. Evolution above the species level. London.

Straka H. 1975. Pollen—und Sporekunde. Stuttgart.

Takhtajan A. 1948. Morphological evolution of the angiosperms. Moscow. (In Russian.)

Takhtajan A. 1959. Die Evolution der Angiospermen. Jena.

Van Campo M. 1971. Précisions nouvelles sur les structures comparées des pollens de Gymnospermes et d'Angiospermes. Comptes Rendus Séances Acad. Sci., Paris (sér. D) 272:2071–2074.

Van Campo M. and B. Lugardon. 1973. Structure grenue infractectale de l'ectexine des pollens de quelques Gymnospermes et Angiospermes. Pollen Spores 15:171–184.

Walker J. W. 1974a. Evolution of exine structure in the pollen of primitive angiosperms. Amer. J. Bot. 61:891–902.

Walker J. W. 1974b. Aperture evolution in the pollen of primitive angiosperms. Amer. J. Bot. 61:1112–1137.

Walker J. W. 1976a. Comparative pollen morphology and phylogeny of the Ranalean complex. In C. B. Beck, ed., Origin and early evolution of angiosperms, pp. 241–299. New York.

Walker J. W. 1976b. Evolutionary significance of the exine in the pollen of primitive angiosperms. In I. K. Ferguson and J. Muller, eds., The evolutionary significance of the exine, pp. 251–308. Linn. Soc. Symp. Series 1. London.

Walker J. W. and J. A. Doyle. 1975. The bases of angiosperm phylogeny: Palynology. Ann. Missouri Bot. Gard. 62:664–723.

Walker J. W. and J. J. Skvarla. 1975. Primitively columellaless pollen: a new concept in the evolutionary morphology of angiosperms. Science 187:445–447.

Walker J. W. and A. G. Walker. 1984. Ultrastructure of Lower Cretaceous angiosperm pollen and the origin and early evolution of flowering plants. Ann. Missouri Bot. Gard. 71:464–521.

Wodehouse R. P. 1935. Pollen grains. New York.

Wodehouse R. P. 1936. Evolution of pollen grains. Bot. Rev. 2:67–89.

Yakovlev M. S. (ed.) 1981. Comparative embryology of flowering plants Winteraceae—Juglandaceae. Leningrad. (In Russian.)

Zavada M. S. 1984. Angiosperm origins and evolution based on dispersed fossil pollen ultrastructure. Ann. Missouri Bot. Gard. 71:444–463.

4

The Ovule, Megasporangium, and Megaspores

The flowering plants inherited from their gymnospermous ancestors the most characteristic organ of all seed plants—the ovule. But the ovules of flowering plants protected by megasporophylls (carpels) underwent a number of significant changes. While in gymnospermous plants, massive ovules with thick envelopes and with comparatively larger reserve of nutrients generally dominate, in magnoliophytes, the ovules are small, tiny, usually with weakly developed envelopes, and generally completely devoid of reserve nutritive substances. The economy of material used for construction of the ovule permitted the flowering plants to increase seed productivity significantly. At the same time, it led to much simplification of the ovule and consequently to its much quicker development than in gymnospermous plants.

The ovule consists of the megasporangium ("nucellus") and the integument, a special protective envelope surrounding it. The integumentary borders leave at the apex of the megasporangium an open narrow canal, a micropyle, through which the pollen tube enters. Typically the ovule is attached to the placenta by means of a more or less apparent stalk, a funicule, although it may be sessile.

4.1. Origin of the Integument

Although the morphological character of the "nucellus" was revealed in the middle of the last century by Hofmeister (1851), who showed that it is a megasporangium, the origin of the integument raises disputes even now. Its morphological character could not be explained for long, till Margaret Benson (1904, Benson and Welsford 1909) put forward her "synangial hypothesis" of the integument of the primitive Palaeozoic gymnosperms.

According to Benson's hypothesis, the ovule originated from the megasynangium. The central megasporangium of this synangium preserved its function and was transformed into the nucellus, while the peripheral megasporangia sterilized and fused together forming the integument. Consequently, a division of function occurred between the central megasporangium—keeping the function of megasporogenesis—and the peripheral sporangia transformed into a protective envelope. According to Benson, each longitudinal chamber of the primitive multichambered Palaeozoic integuments corresponds to one sterilized sporangium. Later, Benson's hypothesis was accepted by a number of morphologists and palaeobotanists, including Kozo-Poljanski (1928), Halle (1933), and Thomas (1936). The teratological material on the ovules of the conifers (Doyle and O'Leary 1934) was cited as confirming it. But the best confirmation of the synangial hypothesis is provided by the primitive ovules of seed ferns, in many cases preserving fairly clear traces of their origin. In the most primitive Upper Devonian and Carboniferous seeds, the integument was segmented (as in *Lagenostoma*), lobed (as in *Archaeosperma, Eurystoma,* and *Physostoma*), or even consisted of more or less separate elongated structures (as in *Moresnetia* and *Genomosperma*). In the Lower Carboniferous *Stamnostoma,* there is already a well-organized integument with micropyle (for the literature, see Stewart 1983). The segmented integuments are found also in the ovules of a number of other seed ferns. It is, however, interesting that they are known not only from the Paleozoic seed ferns but also from the Mesozoic Cycadeoideales and even in the recent Cycadales. Thus, the thick integument of *Cycadeoidea morieri* was divided by radial walls of flattened cells into four longitudinal chambers; and around the micropyle of *Macrozamia, Ceratozamia, Encephalartos,* and other cycads seven to sixteen clear lobes of sclerotesta are visible (De Haan 1920), which correspond to the apices of the integument.

Benson's hypothesis underwent certain changes in connection with the extension of the telome theory to plant morphology. The telomic variant of the synangial hypothesis was proposed by Walton (1940, 1953), Kozo-Poljanski (1948), Pant (1966), and others. According to Kozo-Poljanski (1948), the morphology of a number of Lower Devonian plants like *Yarravia, Hedeia,* and others confirms the possibility of the origin of the ovule directly from the compact "bundle" of simplified telomes. Its origin is related to the shortening of the mesomes. It is therefore quite possible that the integument owed its origin to the coalescence of the belt of sterilized telomes. Similar views were also expressed by Florin (1951), Zimmermann (1959), Andrews (1961, 1963), Camp and Hubbard (1963), Long (1966), Pettit (1970), and others. The telomec theory of the origin of the ovule is a modernized version of Benson's hypothesis.

4.2. Origin of the "Double" Integument

The morphological interpretation of the integument in the magnoliophytes is complicated by the fact that many magnoliopsids and a majority of liliopsids are bitegmic, that is, they have two integuments. In all probability, the outer integument of the angiosperm ovule emerged from the cupule of the ancient gymnospermous ancestor. The cupule is known to have emerged first in the Lyginopteridaceae, but it is not found in these archaic Carboniferous gymnopsperms only. In a modified form, it was also preserved in several later gymnosperms. Already, Mary Stopes (1905) had considered the outer layer of the seed of Cycadaceae or the sacrotesta as a structure homologous to the "outer integument" (i.e., cupule) of *Lagenostoma*. The homology of the "outer integument" and the cupule is still more clearly visible in the Medullosaceae (Takhtajan 1950; Walton 1953). But the cupule gave rise not only to the outer layer of the ovular envelope in a number of gymnospermous seed plants, but also to the outer integument of the magnoliophytes. Some confirmation of this conjecture mentioned by Stebbins (1974:232) is the fact that in many families of flowering plants —including the relatively archaic groups—the outer and inner integuments of the ovule differ greatly from each other in their morphology and their histological structure. In these forms, the outer integument is thicker than the inner one and has specialized epidermal cells, in some cases including stomata. Moreover, the micropyle may be differently shaped in the two integuments. Stebbins mentions also the lobed distal portion of the outer integument in a few genera. Lobed integuments have been observed in Berberidaceae, Juglandaceae, Rosaceae, and Flacourtiaceae (van Heel 1970, 1976). Distal lobing may involve either the outer or the inner integument, or both. "The lobing suggests that the integuments are compound organs," states Bouman (1984:144).

The ovules of flowering plants are either bitegmic (with two integuments) or unitegmic (with one integument); much more rarely, they are ategmic (without integuments). It is generally accepted that unitegmic ovules arose from the bitegmic ones in various lines of evolution (Hallier 1901, 1912; Coulter and

Chamberlain 1903; Maheshwari 1950; Eames 1961; and many others). The most archaic subclass Magnoliidae usually has bitegmic ovules and only rarely are their ovules unitegmic, as in Peperomiaceae, Hydnoraceae, a majority of Rafflesiales, *Ceratophyllum*, a part of Ranunculaceae, Circaeasteraceae, certain Menispermaceae. But all these unitegmic taxa are relatively advanced within the Magnoliidae and Ranunculidae. Unitegmic ovules occur mostly in relatively advanced taxa of magnoliopsids including Ericales, Gentianales, Solanales, Convolvulales, Polemoniales, Boraginales, Scrophulariales, Lamiales, Campanulales, and Asterales, and are rarely found in liliopsids. As the single integument of the sympetalous magnoliopsids (except for Plumbaginales, Primulales, and Cucurbitales) and some choripetalous ones is usually as massive or even more massive than the double, a suggestion was made (Coulter and Chamberlain 1903), that the single massive envelope has a dual character and resulted from the complete fusion of two integuments at the earliest stages of the development of the integumentary primordia. Presumably, in many cases, the unitegmic ovule resulted from the congenital fusion of the integumentary primordia, but in certain taxa it was formed due to abortion of the inner or outer integument. Thus, in *Filipendula, Rubus, Rosa, Potentialla, Fragaria, Alchemilla*, and some other Rosaceae, the envelope resulted from the incomplete development of the inner integument; while in *Peperomia*, Hydnoraceae, *Rafflesia, Mitrastemon, Cytinus*, and some others, it resulted from the underdevelopment of the outer integument. Another pathway of the origin of unitegmy, integumentary shifting, has been described in Ranunculaceae (Bouman and Calis 1977). According to Bouman (1984:140), integumentary shifting is a complicated ontogenetic process involving 1) a fusion of primordia, in the sense that the initials of the two integuments give rise to a common structure; 2) a shifting of the inner integument; and 3) an arrested growth of the latter.

In some families, such as Ranunculaceae, Fagaceae, Betulaceae, Salicaceae, Saxifragaceae, Rosaceae, Styracaceae, Myrtaceae, Fabaceae, Olacaceae, Amaryllidaceae, Orchidaceae, and Poaceae, both bitegmic and unitegmic taxa are found. In some of them, even quite close genera and even species within one genus (*Populus*) are often distinguished by the number of integuments. This shows that the unitegmic condition arose independently and heterochronously in different evolutionary lines.

In some taxa, as a result of reduction, the ovular envelope disappears completely and thus the megasporangium seems to be naked. A good example of an ategmic ovule is provided by *Crinum* (Amaryllidaceae). But as Bouman (1984:151) points out, in most cases the ovular reduction is a complex process involving nucellar reduction, integumentary reduction, and reductions of the funicular and raphal tissues combined with the loss of anatropous curvature. This complex process of reduction is best represented in the order Santalales. The reduction already starts in the relatively archaic family Olacaceae and attains its extreme expression in the families Loranthaceae and Viscaceae. The order Balanophorales is another example of the extreme ovular reduction.

4.3. Form and Orientation of Ovules

In form and orientation, the angiosperm ovules may be of several types, which, however, change into each other. The ovules are formed either with straight nucelli (anatropous, hemitropous, and orthotropous ones) or with curved nucelli (campylotropous and amphitropous ones).

If the axis of the ovule and its funicle are in a straight line, the ovule is termed orthotropous or atropous. Such for example are the ovules of *Lactoris,* Chloranthaceae, Saururaceae, Piperaceae, Peperomiaceae, Hydnoraceae, Cytinaceae *sensu stricto,*

Barclaya, Ceratophyllum, some Lardizabalaceae, Polygonaceae, Altingiaceae, Platanaceae, Myricaceae, Juglandaceae, most of the Cistaceae, Cecropiaceae, Urticaceae, Maundiaceae, Posidoniaceae, Zosteraceae, Cymodoceaceae, and some others.

The anatropous ovule is inverted (bent downward) in its orientation as a result of the approximately 180° curvature of the elongated funicle. It is therefore adherent by its side to the funicle, and the micropyle faces down toward the placenta. In anatropous ovules, the outer (or single) integument is less strongly developed at the funicular side and in some cases is even completely absent. The anatropous curvature, which starts shortly after the initiation of the ovule, is, in fact, an intercalary growth of the funicle (Bouman 1984). Anatropous ovules characterize a majority of flowering plants, including Magnoliales, and are presumably the initial type (Netolitzky 1926; Takhtajan 1959; Eames 1961; Cronquist 1968, 1988; Stebbins 1974; Corner 1976).* If the nucellus forms an angle of about 90° with the funicle, as, for example, in certain Caryophyllaceae, Primulaceae, Scrophulariaceae, Alliaceae, and others, they are called hemitropous (semianatropous).

A special type of the anatropous ovule called circinotropous is found in certain Cactaceae, Amaranthaceae, Chenopodiaceae, and Plumbaginaceae. Due to an intensive one-sided growth, the funicle exceeds 180° and encircles the ovule.

In those cases where the nucellus becomes curved, kidney-shaped, due to an intensive growth on one side, it is called campylotropous. The campylotropous ovule characterizes, for example, many Caryophyllales and Capparales. According to Bouman (1984:147), curvature of the embryo sac is clearly of a

*Gaussen (1946) and Stebbins (1974) explain the primitiveness of anatropous ovules by their origin from the inverted cupules of the Mesozoic seed ferns, particularly the "Corystospermaceae" (Umkomasiaceae) with their one-seeded cupule having "micropylar" tube. "The position of these cupules relative to their stalk was strikingly similar to that of an anatropous ovule having a well-developed funiculus" (Stebbins 1974:232).

derived character, and has the advantage that the embryo may become twice as long as the seed. As it was shown by Bocquet (1959), the campylotropous ovules evolved both from anatropous and the orthotropous. For phylogenetic studies, it is very important to know from which initial type the campylotropous form emerged. In those cases where the origin of the campylotropous ovule is explained we shall be able to refer it to one of the types established by Bocquet—the anacampylotropous or the orthocampylotropous.

If the ovule is bent in the middle so that the nucellus is horseshoe-shaped in longitudinal section, the amphitropous ovule results. The amphitropous ovule characterizes, for example, certain Fabaceae.

4.4. Evolution of the Megasporangium: The Megaspore

The evolution of sporangia in higher plants is characterized by a gradual decrease in the thickness of their walls. The more primitive sporangia with the massive multilayered walls gave rise to more advanced sporangia with thin walls consisting of one layer of cells. In flowering plants, we have both types of megasporangia (nucelli) and, accordingly, two types of ovules —crassinucellate and tenuinucellate (Van Tieghem 1898; Maheshwari 1950; Bouman 1984). In the first type, there is a well-developed parietal tissue and the megaspore mother cell is separated from the nucellar epidermis by one or several parietal layers. In the second type, parietal cells are absent and the megaspore mother cell lies directly below the nucellar epidermis.

The ovules of the overwhelming majority of choripetalous magnoliopsids and of a majority of the liliopsids are crassinucellate. But in sympetalous magnoliopsids and some liliopsids, the ovules are tenuinucellate. These two types are not always strictly exclusive of each other and various transitions between them are known. The tenuinucellate ovule evolved from the crassinucellate due to a reduction of the megasporangial wall (Hallier 1901, 1912; Warming 1913; Netolitzky 1926; Dahlgren 1927; Schnarf 1931; Goebel 1933; Eames 1961; and many others). The most primitive ovules of flowering plants are crassinucellate and bitegmic. Such ovules mainly characterize the choripetalous magnoliopsids and a majority of liliopsids.

In many ovules, where the nucellus disorganizes at an early developmental stage and the female gametophyte comes in direct contact with the integument, the inner layer of the latter becomes specially differentiated and forms a new limiting layer. This layer of integumentary cells is known as the endothelium or integumentary tapetum. The endothelium cells are radially stretched and contain prominent nuclei which can become bi-multinucleate or polyploid. This peculiar tissue has been recorded mainly in families with tenuinucellate or weakly crassinucellate ovules. The endothelium is assumed to be the nutritive layer, whose main function is to act as an intermediary in the transport of nutrition from the integument to the female gametophyte. The presence of this highly specialized tissue characterizes almost all sympetalous magnoliopsids.

At the primordial stage of the ovular development, one or several internal cells near the apex of the nucellus enlarge and become different from the adjacent nucellar cells. They are potential megaspore mother cells or megasporocytes. But, most frequently, only a single functional megasporocyte is formed. It divides meiotically and gives rise to four haploid megaspores.

References

Andrews H. N. 1961. Studies in paleobotany. New York and London.
Andrews H. N. 1963. Early seed plants. Science 142:925–931.

Benson M. J. 1904. *Telangium scottii*, a new species of *Telangium* (*Calymmatotheca*) showing structure. Ann. Bot. 13:161–177.

Benson M. J. and E. J. Welsford. 1909. The morphology of the ovule and female flower of *Juglans regia* and a few allied genera. Ann. Bot. 23:623–633.

Bocquet G. 1959. The campylotropous ovule. Phytomorphology 9:222–227.

Bouman F. 1984. The ovule. In B. M. Johri ed., Embryology of angiosperms, pp. 123–157. Berlin etc.

Bouman F. and J. I. M. Calis. 1977. Integumentary shifting—A third way to unitegmy. Ber. Deutsch. Bot. Ges. 90:15–28.

Camp W. H. and M. M. Hubbard. 1963. On the origin of the ovule and cupule in Lyginopterid pteridosperms. Amer. J. Bot. 50:235–243.

Corner E. J. H. 1976. The seeds of dicotyledons. vols. 1, 2. Cambridge.

Coulter J. M. and C. J. Chamberlain. 1903. Morphology of the angiosperms. New York.

Cronquist A. 1968. The evolution and classification of flowering plants. London.

Cronquist A. 1988. The evolution and classification of flowering plants. 2d ed. New York.

Dahlgren K. V. O. 1927. Die Befruchtungserscheinungen der Angiospermen. Hereditas 10:169–229.

De Haan H. R. M. 1920. Contribution to the knowledge of morphological value and phylogeny of ovule and its integuments. Rev. Trav. Bot. Néerl. 17:219–324.

Doyle J. and M. O'Leary. 1934. Abnormal cones of *Fitzroya* and their bearing on the nature of the conifer strobilus. Sci. Proc. Roy. Soc. Dublin 21:23–35.

Eames A. J. 1961. Morphology of the angiosperms. New York.

Florin R. 1951. Evolution in cordaites and conifers. Acta Horti Berg. 15(11):285–388.

Gaussen H. 1946. Les Gymnospermes, actuelles et fossiles. Trav. Lab. Forest. Toulouse. vol. 2, sec. 1.

Goebel K. 1933. Organographie der Pflanzen. III. Jena.

Halle T. G. 1933. The structure of certain fossil sporebearing organs believed to belong to pteridosperms. Kungl. Svenska Vetensk. Acad. Handl. ser. 3, 12:1–103.

Hallier H. 1901 (1902). Beiträge zur Morphogenie der Sporophylle und Trophophylls in Beziehung zur Phylogenie der Kormophyten. Jahrl. Hamburg. Wiss. Anst. 19, 3. Beih.:1–110.

Hallier H. 1912. L'origine et la système phylétique des Angiosperms exposés à l'aide de leur arbre généalogique. Arch. Néerl. Sci. Exact. Nat. ser. 3 B, 1:146–234.

Heel W. A. van. 1970. Distally lobed integuments in some angiosperm ovules. Blumea 18:67–70.

Heel W. A. van. 1976. Distally-lobed integuments in *Exodhorda, Juglans, Leontice,* and *Bongardia.* Phytomorphology 26:1–4.

Hofmeister W. 1851. Vergleichendl Untersuchungen der Keimung, Entfaltung und Fruchtbildung hoherer Kryptogamen und der Samenbildung der Coniferen. Leipzig.

Kozo-Poljanski B. M. 1928. The ancestors of the angiosperms. Moscow. (In Russian.)

Kozo-Poljanski B. M. 1948 (1949). To the modernization of the system of the plant world. Trudy Voronezh State Univ. 15: 76–129. (In Russian.)

Long A. G. 1966. Some lower Carboniferous fructifications from Berwickshire, together with a theoretical account of the evolution of ovules, cupules and carpels. Trans. Roy. Soc. Edinburgh 66:345–375.

Maheshwari P. 1950. An introduction to the embryology of angiosperms. New York.

Maheshwari P. 1960. Evolution of the ovule. 7th Seward Mem. Lecture, Sahni Inst. Palaeobotany, Lucknow, pp. 3–13.

Netolitzky F. 1926. Anatomie der Angiospermen—Samen. In K. Linsbauer, ed., Handbuch der Pflanzenanatomie. Berlin.

Pant D. D. 1966. Origin of ovules or seeds and their integuments. In: T. S. Mahabale, ed., Proceeding of the autumn school in botany, pp. 237–253. Poona (India).

Pettit J. 1970. Heterospory and the origin of the seed habit. Biol. Rev. 45:401–415.

Savchenko M. I. 1973. Morphology of the ovule in angiosperms. Leningrad. (In Russian.)

Schnarf K. 1931. Vergleichende Embryologie der Angiospermen. Berlin.

Sporne K. R. 1969. The ovule as an indicator of evolutionary status in angiosperms. New Phytol. 68:555–566.

Stebbins G. L. 1974. Flowering plants. Evolution above the species level. Cambridge, Mass.

Stewart W. N. 1983. Paleobotany and the evolution of plants. Cambridge.

Stopes M. 1905. On the double nature of the cycadean integument. Ann. Bot. 19:561–566.

Takhtajan A. 1950. Phylogenetic principles of the system of higher plants. Bot. Zhurn. (Leningrad) 35:113–135. (In Russian.) English translation: Bot. Rev. 1953, 19:1–45.

Takhtajan A. 1959. Die Evolution der Angiospermen. Jena.

Thomas H. H. 1936. Palaeobotany and the origin of the angiosperms. Bot. Rev. 2:397–418.

Van Tieghem P. 1898. La structure de quelques ovules et le parti qu'on en peut tirer pour améliorer la classification. J. Bot. (Paris) 12:197–220.

Walton J. 1940. An introduction to the study of fossil plants. London.

Walton J. 1953. The evolution of the ovule in the pteridosperms. Adv. Sci. 10:223–230 (British Association Adv. Sci. no. 38).

Warming E. 1878. De l'ovule. Ann. Sci. Nat. Bot. s/r. 6, 5:177–266.

Warming E. 1913. Observations sur la valeur syst/matique de l'ovule. Mind. Jap. Steenstrup 24:1–45.

Zimmermann W. 1959. Die Phylogenie der Pflanzen. 2 Aufl. Stuttgart.

5

Evolutionary Trends in Pollination

During the evolution of flowering plants the mechanism of pollination evolved in the most diverse directions and, in many cases, attained very great complexity. An overwhelming majority of flowering plants are characterized by cross-pollination of their flowers. Self-pollination has evidently arisen as a secondary phenomenon, and self-pollinating species are usually the ends of evolutionary lines (see especially Stebbins 1957, 1974). According to Stebbins (1974) the shift from obligate outcrossing to self-pollination occurs chiefly, and perhaps exclusively, in species that occupy temporary, pioneer habitats. Self-pollination may be the only way in which the new colonizers can produce offspring rapidly. Therefore, self-pollinated species are often at great advantage in long-distance dispersal (Baker 1955).

Cross-pollination is attained by exceptionally varied methods and through quite diverse carriers. The main vectors of pollen transfer are insects, wind, and birds, more rarely, bats and some wingless mammals and, in some aquatic plants, water currents.

Long ago, the idea was expressed that in angiosperm evolution, entomophily was the initial type of pollination (Henslow 1888; Bessey 1897; Robertson 1904; and others). Robertson (1904) cited quite convincing evidences in favor of the primitiveness of entomophily and the secondary character of all other means of pollen transfer. In particular, he noted that the typical anemophilous flowers bear the imprint of reduction and simplification and most frequently give rise to monospermous indehiscent fruits. The reduction of the number of ovules in the ovary with the transition from entomophily is related, according to Robertson, to the fact that in the second case it is more difficult to entrap a sufficient number of pollen grains for a large number of ovules. In his opinion (Robertson 1917), the stigma and the ovary are themselves entomophilous features and developed after the establishment of entomophily.

According to van der Pijl (1960), entomophily should have taken place before the origin of angiosperms, i.e., already in the ancestors of the latter. Leppik (1960, 1977) suggested that beetles and other insects first became pollen gatherers in association with the Cycadeoideales, and later transferred their activity to the evolving angiosperms.* Some of extant cycads are also pollinated by beetles (see Rattray 1913; and especially Norstog et al. 1986 and Norstog 1987). Most probably, the immediate ancestors of flowering plants were already associated with various insect pollinators.

The initial bait of the flower for attracting the insects was most probably the pollen (Darwin 1876; van der Pijl 1960; Grinfeld 1962; Faegri and van der Pijl 1979; Willemstein 1987; and others), which existed before the origin of flowers. Even now, many archaic flowering plants, as for example, Annonaceae, some Winteraceae, Amborellaceae, Lactoridaceae, Chloranthaceae, Schisandraceae, and Papaveraceae, have no other food

*According to Crepet (1979, 1983), Crepet and Friis (1987), and Gottsberger (1988), fundamental bennettitalean features suggest that the group may have been beetle pollinated. According to Gottsberger (1988:634), *Cycadeoidea*, with its permanently closed strobiles ("flowers"), probably had attained the ability to transform the casual beetle visitors into "faithful" pollinators. The closing of microsporophylls over the strobile he considered an adaptation for beetle pollination.

supply except for pollen. But pollen is too precious for such disorderly expenditure. The necessity for pollen economy leads to the origin of cheaper food supply, staminodes and other food bodies and nectar. Staminodes occur in many archaic taxa, mainly in the Magnoliidae (see Endress 1984b). The next evolutionary stage of the evolution of sterile food supply is the emergence of nonsecretory glandular food bodies in the flowers, like glandular warts on the stigma (Magnoliaceae), food bodies at the apices of the connectives (Calycanthaceae), glands on the staminodes (Austrobaileyaceae), food bodies at the base of some tepals (*Liriodendron*), and some other flower parts including fleshy petals. According to Willemstein (1987) some types of nectaries could probably originate from the glandular bodies. This could happen on the base of the filaments (as in the Fumariaceae), at the carpels (*Caltha*), on the tepals or petals (Berberidaceae and Lardizabalaceae), or on the sepals (the majority of the Ranunculaceae). But, usually, the nectaries developed independently of the food bodies. They orginated in the most diverse lines of the evolution of flowering plants and on most widely varying morphological bases (see especially Brown 1938; Norris 1941; Fahn 1953; Kartashova 1965; Willemstein 1987).

Attraction of pollinators by means of floral odors and colors played an increasing role in the evolution of pollination. Evolution of flower color is of special interest. According to Hildebrand (1879), H. Müller (1883), Allen (1891), Henslow (1888), Willemstein (1987), and others, the first color departing from the primitive green was yellow. In fact, the change from green to yellow only requires the loss of chlorophyll. And when the chlorophyll in autumn leaves disappears, the remaining carotenoids—xanthophyll, which is yellow, and carotene, which is orange, color the leaves yellow or orange. Carotenoids also occur independently of chlorophyll as flower pigments and are characteristic for the pollen of most entomophilous flowers.

They are probably the earliest angiosperm floral pigments (Willemstein 1987). But whereas yellow and red-orange are caused by carotenoids, pink, red, violet, and blue are caused by anthocyanins, which are the major class of flavonoids. The flavonols, another group of flavonoids, are found in many flowers and may contribute to their ivory or white hues, although insects see them differently (see Proctor and Yeo 1973; Dement and Raven 1973, 1974; Faegri and van der Pijl 1979; Barth 1982).

According to Müller (1883), the evolutionary sequence of the colors might have been as follows: greenish-yellow, yellow, yellow with red spots, white, white with yellow spots, rose red, and red. Willemstein (1987:178) gives a more elaborate scheme with four theoretically possible transformation series: 1) from (green) yellow to white: this development could have taken place in the very early evolution of the angiosperm flowers; 2) from yellow or white to blue-violet and blue-mixed colors: this development possibly started in the Late Cretaceous; 3) from yellow or white to red and red-mixed colors: this development possibly also started in the Late Cretaceous; 4) from blue to red or vice versa: these developments might have been functional from the Late Cretaceous onwards.

The reds are a relatively late acquisition, and only lately originated blues, which "seems probable from the comparative rarity of the last color" (Henslow 1888:179).

The evolution of entomophily began with very primitive forms. The first pollinators were not, of course, such specialized suctorial insects as butterflies and bees, but less specialized pollen-feeding and short-tongued pollinators, with short mouth parts (without a proboscis). The orginal pollinators might have been beetles (Diels 1916; Kozo-Poljanski 1922; Grant 1950b; Porsch 1950; Leppik 1957, 1960, 1963; Takhtajan 1959, 1969; van der Pijl 1960, 1961, 1969; Eames 1961; Heiser 1962; Baker 1963; Percival 1965; Baker and Hurd 1968; Stebbins 1974; Faegri and van der Pijl 1979; and others). Since Delpino's (1868–1875) classical work, many observations have been made on beetle pollination in various archaic magnoliids including Magnoliaceae (Heiser 1962; Thien 1974; Gibbs et al. 1977; Gottsberger 1977; 1986); *Degeneria* (Takhtajan 1973, 1980; Thien 1980); *Eupomatia* (Hamilton 1897; Hotchkiss 1958; Endress 1984a); Annonaceae (van der Pijl 1953, 1960; Gottsberger 1970, 1985, 1988; Weber 1981; Endress 1985); *Myristica* (Armstrong and Drummond 1986, Armstrong and Irvine 1989); some Winteraceae (Gottsberger et al. 1980; Thien 1980; Bernhardt and Thien 1987; Gottsberger 1988); *Calycanthus* (Grant 1950a); and *Victoria* (Prance and Arias 1975). In some of them, such as *Degeneria, Eupomatia,* many Annonaceae, and *Calycanthus,* during the female phase the flowers are almost closed ("insect-trapping flowers") and only slightly open at the apex—a characteristic feature of typical cantharophilous flowers. But some others, such as Winteraceae, are not so narrowly specialized and "have a much broader array of pollinators" (Gottsberger 1988:636).

The beetles, which are very ancient (although Nitidulidae, only of the Late Jurassic), were most probably one of the earliest pollinators of the original flowering plants. But they most likely were not the only pollinators of the earliest magnoliophytes.

A pollen-eating moth *Sabatinca,* which belongs to the most ancient lepidopteran family Micropterigidae, pollinates flowers of the archaic genus *Zygogynum* (Winteraceae) and could probably also represent the earliest pollinators (see Thien et al. 1985; Bernhardt and Thien 1987).

Another group of possible early pollinators is Diptera, which are also very ancient (the oldest fossils date the Triassic). Many modern fly families eat pollen (Syrphidae, Calliphoridae, etc.). According to Thien (1980), the early angiosperm flower may have been associated with sugar-feeding Diptera as well as with pollen-eating Coleoptera. He notes that the flowers of the Old

World species of *Drimys* are adapted for pollination by many species of Diptera, although the flowers are occcasionally also visited by Coleoptera and Hymenoptera. A great number of flies are also observed on the flowers of *Liriodendron tulipifera* and *Nymphaea* ssp. (Thien 1980; Meeuse and Schneider 1980), as well as on the flowers of *Illicium floridanum* (Thien, White, and Yatsa 1983) and *Pseudowintera* (Thien et al. 1985).

Plecoptera (stoneflies) and Thysanoptera (thrips) were probably also among the first pollinators. Stoneflies are among the pollinators of *Illicium* (Thien et al. 1983; White and Thien 1986), and thrips pollinate *Belliolum* (Thien 1980), *Drimys* (Gottsberger et al. 1980), and *Pseudowintera* (Godley and Smith 1976). Therefore, together with beetles, micropterigid moths, flies, stoneflies, and thrips are among the most likely pollinators of the earlier flowering plants.

The Russian entomologist Malyshev (1964, 1968) suggested that the first pollinators of flowering plants were some extinct wasplike ancestors of the honeybees (Apoidea). Willemstein (1987) has also suggested that wasps participated in the pollination of the earliest flowering plants (together with the beetles and flies). But wasps "are not interested in pollen" (Faegri and van der Pijl 1979:107), and wasp pollination is not characteristic for the extant archaic magnoliophytes.

At the beginning of the Late Cretaceous or perhaps earlier, the initial primitive pollen-feeding pollinators with short mouth parts—whether they were beetles or insects of some other kind —had gradually faded into the backgound, as the leading role in pollination passed to specialized pollinators. The development of the floral tube during the Cretaceous led to the exclusion of those visitors whose mouth parts were too short to reach the nectar and hence to the evolution of the proboscis, which elongated in parallel with the floral tube. As a result of this coevolution, there arose highly specialized pollinators among

Hymenoptera (long tongued Apoidea), Diptera (flies with longer probosces), Lepidoptera (butterflies and moths), birds, and bats.

The development of specialized insect pollinators led to a very great advance in the foraging strategies of pollinators, including an increase in the constancy of flower visitation (see Grant 1950c; Proctor and Yeo 1973; Levin 1978). Aristotle, in his "History of Animals," had already noted that a bee visited flowers of only one species on each flight, e.g., only flowers of the violet. Darwin (1876), who specially studied this problem, stated that species of bees and some other insects visit the flowers of one and the same species as long as they can obtain food from it before moving to another species. If the population of a particular species is sufficiently numerous, the pollination of the flowers of its individuals may continue for the whole "working day" of the pollinator. In the case of a small population, the insect easily moves over to another species. Although a more or less evident constancy of visiting the same species of plants on each flight is found already in some beetles, it attains in Hymenoptera, Lepidoptera, and Diptera the maximum development (Kugler 1955). The more specialized pollinating insects are characterized by highly developed instincts of flower constancy and by a high capacity of creating new conditioned reflexes. At the same time, the development of the constancy of visiting the flowers by pollinators demanded a considerable progress in the structure of flowers themselves and, above all, a corresponding perfection in the concealment of the nectar and pollen. And what is also very important, it required standardization of the dimensions of the flower and its component parts in line with the dimensions of the body and proboscis of the specific pollinator (see Berg 1956, 1958).

Flowers with many stamens and easily accessible food—pollen or nectar—are marked by the absence or weak development of constancy of visitation. In these cases, the insects often fly from flowers of one species to those of another, and constancy, if it exists at all, is of an incipient nature. This is so for many choripetalous and actinomorphic flowers. Things are quite different in the case of zygomorphic and especially sympetalous flowers. In such specialized flowers, the number of stamens and carpels is reduced. The constancy of visiting is most developed, and there is more opportunity for concealmnent of nectar and pollen. For species with such flowers, definite constancy of visitation by the pollinators—usually insects—has become established. In such flowers, pollination is carried out by bees, moths, butterflies, and long-tongued flies, and in some cases by birds and bats (see Grant 1949; 1950b; Kugler 1955; Percival 1965; Proctor and Yeo 1973; Faegri and van der Pijl 1979). As was shown by Müller and Darwin, flower constancy is mutually beneficial to both plant and pollinator. The former is better pollinated with less expense of pollen and nectar; the latter can obtain food more easily. It represents the highest achievement of the coevolution of flowering plants and their pollinators and is the result of a complex process of mutual selection.

In some lines of angiosperm evolution, a transition has taken place from entomophily to anemophily. The main condition for the transition to wind pollination was probably a scarcity of pollinators. This scarcity of pollinators is particularly noticeable in early spring, in cold and high altitude regions, as well as in deserts and marshes. The transition from entomophily to anemophily led to significant structural and functional transformation of the flower and, above all, to the reduction and atrophy of the corolla, to a considerable decrease in general size of the flower, to special changes in the stigma and stamens, to a separation of sexes, and to the formation of characteristic inflorescences (see Faegri and van der Pijl 1979). Palaeobotanical evidence suggests that specialized wind-pollination evolved not later than in the Albian.

Anemophily resulted from entomophily in diverse lines of

evolution of both magnoliopsids and liliopsids. The anemophilous plants are found among archaic groups (e.g., *Tetracentron*) and among highly advanced orders. In many plants where anemophily appeared relatively not very long time ago, there are more or less close entomophilous relatives. Such, for example, is the case with *Macleaya, Rumex, Poterium, Sanguisorba, Ricinus, Acer, Fraxinus, Ambrosia, Juncaceae,* and others, while *Thalictrum* is still partly entomophilous. But even where the specialization towards anemophily went so far that the entire family has no more entomophilous forms in it, the origin from entomophilous ancestors does not raise any doubt. Thus, it is fully evident that the anemophilous families Casuarinaceae, Fagaceae, Betulaceae, Myricaceae, and Juglandaceae emerged from the entomophilous ancestors, most probably from Hamamelidales. The evolution of the order Hamamelidales itself proceeded from entomophily to anemophily, which is here expressed in the gradual development of capitate or catkinlike inflorescences, simplification of flowers, diminution in the number of carpels, reduction of the perianth, and tendency to apetaly. The origin of anemophilous Potamogetonales from entomophilous Alismatales, of anemophilous Juncaceae and Cyperaceae from entomophilous Lilianae, or the origin of grasses from the entomophilous ancestors is no less clear among liliopsids.

In a number of cases, a phenomenon is observed where there is a return to entomophily in groups earlier adapted to wind pollination. Characteristic examples of this kind are species of *Dichromena* (Cyperaceae), such as *D .ciliata* and *D. latifolia* with their specialized inflorescences. *Castanea,* Nyctaginaceae, some Chenopodiaceae, *Salix,* and some Araceae can also serve as example of plants which have returned to entomophily.

References

Allen G. 1891. The colours of flowers as illustrated by the British flora. London.

Armstrong J. A. and B. A. Drummond III. 1986. Floral biology of *Myristica fragrans* Houtt. (Myristicaceae), the nutmeg of commerce. Biotropica 18:32–38.

Armstrong J. A. and A. K. Irvine. 1989. Floral biology of *Myristica insipida* (Myristicaceae), a distinctive beetle pollination syndrome. Amer. J. Bot. 76(1):86–94.

Baker H. G. 1955. Self-compatibility and establishment after "long-distance" dispersal. Evolution 9:347–348.

Baker H. G. 1963. Evolutionary mechanisms in pollination biology. Science *139* (3558):877–883.

Baker H. G. and P. D. Hurd. 1968. Introfloral ecology. Ann. Review Entom. 13:385–414.

Barth F. G. 1982. Biologie einer Begegnung. Die Partnerschaft der Insekten und Blumen. Stuttgart. (English transl., F. G. Barth 1985. Insects and flowers. The biology of partnership. Princeton.)

Berg R. L. 1956. The standardizing selection in the evolution of the flower. Bot. Zhurn. 41:318–334 (In Russian.)

Berg R. L. 1958. Further investigation in the stabilizing selection in the evolution of flowers. Bot Zhurn. 43:12–27 (In Russian.)

Bernhardt P. and L. B. Thien. 1987. Self-isolation and insect pollination in the primitive angiosperms: new evaluations of older hypotheses. Pl. Syst. Evol. 156:159–176.

Bessey C. E. 1897. Phylogeny and taxonomy of the angiosperms. Bot. Gaz. 24:145–178.

Brown W. H. 1938. The bearing of nectaries on the phylogeny of flowering plants. Proc. Amer. Phil. Soc. 79:549–595.

Crepet W. L. 1979. Insect pollination: a paleontological perspective. BioScience 29:102–108.

Crepet W. L. 1983. The role of insect pollination in the evolution of the angiosperms. In L. Real, ed., Pollination biology, pp. 29–50. Orlando.

Crepet W. L. and E. M. Friis. 1987. The evolution of insect pollination in angiosperms. In E. M. Friis, W. G. Chaloner, and P. R. Crane, eds., The origins of angiosperms and their biological consequences, pp. 181–201. Cambridge.

Cronquist A. 1968. The evolution and classification of flowering plants. New York.

Darwin C. 1876. The effects of cross- and self-fertilization in the vegetable kingdom. London.

Daumann E. 1930. Das Blütennektarium von *Magnolia* und die Fütterkörper in der Blüte von *Calycanthus*. Planta 11:108–116.

Delpino F. 1868–1875. Ulteriori osservazioni sulla dicogamia nel regno vegetale, I-II. Atti Soc. Ital. Sc. Nat., Milano, pp. 11–12.

Dement W. A. and P. H. Raven. 1973. Distribution of the chalcone, isosalipurposide, in the Onagraceae. Phytochemistry 12:807–808.

Dement W. A. and P. H. Raven. 1974. Pigments responsible for ultraviolet patterns in flower of *Oenothera* (Onagraceae). Nature 252:705–706.

Diels L. 1916. Käferblumen bei den Ranales und ihre Bedeutung für die Phylogenie der Angiospermen. Ber. Dt. Bot. Ges. 34:758–774.

Eames A. J. 1961. Morphology of the angiosperms. New York, Toronto, and London.

Endress P. K. 1984a. The flowering process in the Eupomatiaceae (Magnoliales). Bot. Jahrl. Syst. 104:297–319.

Endress P. K. 1984b. The role of inner staminodes in the floral display of some relic Magnoliales. Plant Syst. Evol. 146:269–282.

Endress P. K. 1985. Stamenabszission und Pollenpräsantation bei Annonaceae. Flora 176:95–98.

Faegri K. and van der Pijl L. 1979. The principles of pollination ecology. 3d ed. Oxford.

Fahn A. 1953. The topography of the nectary in the flower and its phylogenetic trend. Phytomorphology 3(4):424–426.

Friis E. M. and W. L. Crepet. 1987. Time appearance of floral features. In E. M. Friis, W. G. Chaloner, and P. R. Crane, eds., The origins of angiosperms and their biological consequences, pp. 145–179. Cambridge.

Gibbs P.E., J. Semir, and N. D. da Cruz. 1977. Floral biology of *Talauma ovata* St. Hill. (Magnoliaceae). Ciênc. Cult. 29:1436–1441.

Godley E. J. and D. H. Smith. 1976. Self-sterility in *Pseudowintera colorata*. Triennial report (Botany Division, DSIR) 75, 61.

Gottsberger G. 1970. Beiträge zur Biologie von Annonaceen-Blüten. Ost. Bot. Z. 118:237–279.

Gottsberger G. 1974. The structure and function of the primitive Angiosperm flower—a discussion. Acta Bot. Neerl. 23(4):461–471.

Gottsberger G. 1977. Some aspects of beetle pollination in the evolution of flowering plants. Plant Syst. Evol., Suppl. 1:211–226.

Gottsberger G. 1985. Pollination and dispersal in the Annonaceae. Annonaceae Newslett. (Utrecht) 1:6–7.

Gottsberger G. 1986. Some pollination strategies in neotropical savannas and forests. Plant Syst. Evol. 152:29–45.

Gottsberger G. 1988. The reproductive biology of primitive angiosperms. Taxon 37(3):630–643.

Gottsberger G., I. Silberbauer-Gottsberger, and F. Ehrendorfer. 1980. Reproductive biology in the primitive relic angiosperm *Drimys braziliensis* (Winteraceae). Plant Syst. Evol. 135:11–39.

Grant V. 1949. Pollination systems as isolating mechanisms in angiosperms. Evolution 3:82–97.

Grant V. 1950a. The protection of the ovules in flowering plants. Evolution 4(3):179–201.

Grant V. 1950b. The pollination of *Calycanthus occidentalis*. Amer. J. Bot. 37:194–297.

Grant V. 1950c. The flower constancy of bees. Bot. Rev. 16:379–398.

Grinfeld E. K. 1962. The origin of anthophily in insects. Leningrad. (In Russian.)

Hamilton A. G. 1897. On the fertilization of *Eupomatia laurina* R. Br. Proc. Linn. Soc. N. S. Wales 22:48–55.

Heiser C. B. 1962. Some observations on pollination and compatibility in *Magnolia*. Proc. Indiana Acad. Sci. 72:259–266.

Henslow G. 1888. The origin of floral structures through insect and other agencies. London.

Heslop-Harrison Y. and K. R. Shivanna. 1977. The receptive surface of the angiosperm stigma. Ann. Bot. 41:1233–1258.

Hildebrand F. 1879. Die Farben der Blüten. Leipzig.

Hotchkiss A. T. 1958. Pollen and pollination in the Eupomatiaceae. Proc. Linn. Soc. N. S. Wales 83:86–91.

Kartashova N. N. 1965. The structure and function of nectaries of dicotyledonous plants. Tomsk. (In Russian.)

Kozo-Poljanski B. M. 1922. An introduction to phylogenetic systematics of the higher plants. Voronezh. (In Russian.)

Kugler H. 1955. Einführung in die Blütenökologie. Jena.

Leppik E. E. 1957. Evolutionary relationship between entomophilous plants and anthophilous insects. Evolution 11:466–481.

Leppik E. E. 1960. Early evolution of flower types. Lloydia 3:72–92.

Leppik E. E. 1963. Fossil evidence of floral evolution. Lloydia 26:91–115.

Leppik E. E. 1977. Floral evolution in relation to pollination ecology. New Delhi.

Levin D. A. 1978. Pollinator behaviour and the breeding structure of plant populations. In A. J. Richards, ed., The pollination of flowers by insects, pp. 133–150. Linnean Soc. Symposium Series. no. 6. London.

Malyshev S. I. 1964. Formation of flowering plants in connection with the behavioural evolution of waspoid ancestors in Apoidea. Advances of modern biology. 57(1):159–174. (In Russian.)

Malyshev S. I. 1968. Genesis of the Hymenoptera and the phases of their evolution. Methuen, London.

Meeuse B. J. D. and E. L. Schneider. 1980. *Nymphaeae* revisited: a preliminary communication. Israel J. Bot. 28:65–79.

Müller H. 1883. Die biologische Bedenutung der Blumenfarben. Biol. Centralb. 3:97.

Norris Th. 1941. Torus anatomy and nectary characteristic as phylogenetic criteria in the Rhoeadales. Amer. J. Bot. 28:101–113.

Norstog K. J. 1987. Cycads and the origin of insect pollination. American Scientist 75:270–279.

Norstog K.J., D. W. Stevenson, and K. J. Niklas. 1986. The role of beetles in the pollination of *Zamia furfuracea* L. fil. (Zamiaceae). Biotropica 18(4):300–306.

Pellmyr O. and L. B. Thien. 1986. Insect reproduction and floral fragrances: keys to the evolution of the angiosperms. Taxon 35(1):76–85.

Percival M. 1965. Floral biology. Elmsford, N.Y.

Pijl L. van der. 1953. On the flower biology of some plants from Java. Ann. bogot. 1:77–99.

Pijl L. van der. 1960, 1961. Ecological aspects of flower evolution. I-II. Evolution. 14:403–416, 15:44–59.

Pijl L. van der. 1969. Evolutionary action of tropical animals on the reproduction of plants. Biol. J. Linn. Soc. 1:85–82.

Porsch O. 1950. Geschichtliche Lebenswertung der Kastanienblüte. Ost. Bot. Z. 97:359–372.

Prance G. G. and A. B. Anderson. 1976. Studies of the floral biology of neotropical Nymphaeaceae 3. Acta Amazonica 5:109–139.

Prance G. G. and J. R. Arias. 1975. A study of the floral biology of *Victoria amazonica* (Poepp.) Sowerby (Nymphaeaceae). Acta Amazonica. 5:109–139.

Proctor M. and P. Yeo. 1973. The pollination of flowers. London.

Rattray G. 1913. Notes on the pollination of some South African cycads. Trans. R. Soc. S. Afr. 3:253–270.

Robertson C. 1904. The structure of the flowers and the mode of pollination of the primitive angiosperms. Bot. Gaz. 37:294–298.

Robertson C. 1917. Flowers and insects. XX. Evolution of entomophilous flowers. Bot. Gaz. 58:307–316.

Stebbins, G. L. 1957. Self-fertilization and population variability in the higher plants. Amer. Natur. 91:337–354.

Stebbins G. L. 1974. Flowering plants. Evolution above the species level. Cambridge, Mass.

Takhtajan A. 1959. Die Evolution der Angiospermen. Jena.

Takhtajan A. 1969. Flowering plants. Origin and dispersal. Edinburgh.

Takhtajan A. 1973. Evolution und Ausbreitung der Blütenpflanzen. Jena.

Takhtajan A. 1980. Degeneriaceae. In A. Takhtajan, ed., The life of plants, 5(1):121–125. (In Russian.)

Thien L. B. 1974. Floral biology of *Magnolia*. Amer. J. Bot. 61:1037–1045.

Thien L. B. 1980. Patterns of pollination in the primitive angiosperms. Biotropica 12(1):1–13.

Thien L. B., P. Bernhardt, G. W. Gibbs, O. Pellmyr, Groth I. Bergström, and G. McPherson. 1985. The pollination of *Zygogynum* (Winteraceae) by a moth, *Sabatinca* (Micropterigidae): an ancient association? Science 227:540–542.

Thien L. B., D. A. White, and L. Y. Yatsa. 1983. The reproductive biology of a relict—*Illicium floridanum* Ellis. Amer. J. Bot. 70:719–727.

Weber A. C. 1981. Alguns aspectos da biologia floral de *Annona sericea* Dun. (Annonaceae) Acta Amazonica 11:61–65.

White D. A. and L. B. Thien. 1986. The pollination of *Illicium parviflorum* Michx. ex Vent (Illiciaceae). J. Elisha Mitchell Soc.

Willemstein S. C. 1987. An evolutionary basis for pollination biology. Leiden.

6

Evolution of Male and Female Gametophytes: Fertilization and Triple Fusion

In the course of evolution, both the male and female gametophytes of flowering plants reached a very high degree of simplification and specialization. Gametogenesis occurs in them at such an early stage of an extremely abbreviated ontogeny of the gametophyte that gametangia cannot even be formed, and the gametes are formed without them. Moreover, the development of the gametes themselves is also cut short, and they became extremely simplified. Due to a sharp abbreviation and acceleration of their ontogeny, the gametophytes of magnoliophytes completely lost their gametangia. As I have suggested in my previous works (beginning in 1948) these drastic changes in the gametophyte structure and development resulted from neoteny and subsequent specialization (see especially Takhtajan 1959 and 1976).

As a result of fundamental modification of ontogeny, there appeared in flowering plants (as well as in a part of the gymnosperms) completely new types of "gametangialess" gametophytes and new types of gametes sharply different from the mature gametophytes and gametes of the Cycadophyta and other "archegoniate" gymnospermous plants. But it should be noted that the general evolutionary trend towards an abbreviation and simplification of ontogeny and acceleration of the gametogenesis started already among gymnospermous plants.

6.1. Origin of the Male Gametophyte

The entire male gametophyte of the flowering plants consists only of two cells—a smaller generative cell lying next to the microspore wall and a large tube cell ("vegetative" cell). In contrast to male gametophytes of a majority of gymnospermous plants, it has neither the prothallial cells, nor the stalk cell and the true spermatogenous cell ("body cell"). The function of the spermatogenous cell has been transferred to the generative cell, which divides to form two nonmotile male gametes, and the function of the stalk cell ("dislocator") became unnecessary. Thus the angiosperm male gametophyte reached the climax of simplification and miniaturization, which precluded any further major structural changes.

The reduction and simplification of the angiosperm male gametophyte occurred through the loss of both the late stages (terminal abbreviation) and the initial ones (basal abbreviation). If the ontogeny and the number of mitotic divisions of the male gametophyte of *Pinus* and flowering plants are compared, it becomes clear that the first two divisions as well as the fourth (the last in the ontogeny of the male gametophyte of *Pinus*) fell out from the development of the angiosperm male gametophyte as a result of abbreviations. Thus, in the development of the male gametophyte of flowering plants, there are neither prothallial cells (figure 20—formed in *Pinus* as a result of the first two divisions and soon flattened and dead), nor stalk cell (sterile cell) and the spermatogenous cell (body cell) (resulting from the fourth division).

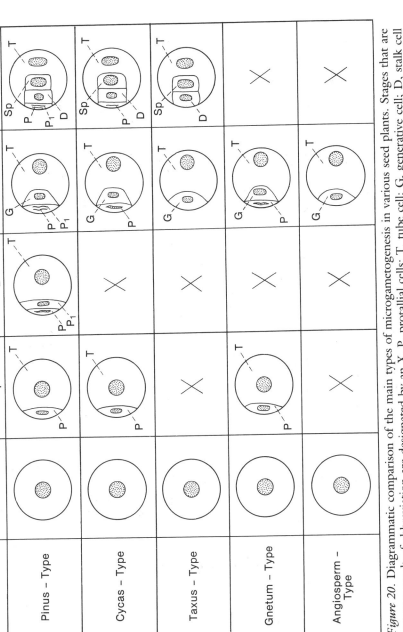

Figure 20. Diagrammatic comparison of the main types of microgametogenesis in various seed plants. Stages that are lost as a result of abbreviation are designated by an X. P, protallial cells; T, tube cell; G, generative cell; D, stalk cell (dislocator); S, spermatogeneous cell (from A. Takhtajan 1959).

The male gametes in flowering plants also undergo radical transformation and constitute a completely new structure, which replaces both the stalk and spermatogenous cells of the ancestral gametophyte. The stalk cell and the spermatogenous cell are the only remnants of an antheridium in gymnospermous plants (except in *Gnetum* and *Welwitshia,* which have lost both of these cells). In a certain sense, we can even say that the stalk cell and spermatogenous cell were transformed into male gametes of *Welwitschia, Gnetum,* and flowering plants (Takhtajan 1976).

In the majority of flowering plants, including the archaic taxa, the pollen is released from the anther in the two-celled stage in the development of the gametophyte. But in many other flowering plants, including some advanced taxa, the generative cell divides before the pollen grain is shed and the male gametophyte is therefore three-celled. "This early division of the generative nucleus apparently represents one more step in the progressive compression of the gametophyte that characterizes vascular plants in general," states Cronquist (1988:215). Therefore the two-celled condition is primitive and the three-celled type is derived and orginated independently in many lines of angiosperm evolution (see Schnarf 1939; Rudenko 1959; Brewbaker 1967).

6.2. Origin of the Female Gametophyte

The female gametophyte (embryo-sac) of flowering plants is considerably less reduced and simplified than the male gametophyte and is therefore more liable to evolutionary modifications, both of development and of structure. The different types of female gametophytes are distinguished mainly on the basis of the number of megaspores or megaspore nuclei that participate in their formation, on the number of mitotic divisions during gametogenesis, and the number and arrangement of the cells and free nuclei present in the mature gametophyte (see Johri 1963; Romanov 1971; Willemse and van Went 1984).

On the basis of the number of megaspores or megaspore nuclei participating in the formation of female gametophyte, three major types of gametophytes are recognized—monosporic, bisporic, and tetrasporic.

In most flowering plants, the female gametophyte is monosporic, developing from one of the four potential megaspores separated by cell walls. The three upper ones degenerate gradually, and the lowermost (chalazal) develops into the female gametophyte as a result of three successive mitotic divisions. The first division of the nucleus of the functioning megaspore gives rise to two nuclei that move apart to the poles of a much elongated megaspore. The successive divisions of these nuclei at each pole yield eight nuclei arranged in groups of four at the micropylar and chalazal ends of the female gametophyte. Subsequently, one nucleus from each of the poles migrates to the central region of the gametophyte. These so-called polar nuclei converge and may fuse—either immediately or only before fertilization—to form a diploid secondary nucleus, but at times they may remain separate until fertilization. Three of the four nuclei of each pole are covered with cytoplasm and thus transformed into cells tightly in contact with each other. Those at the micropylar end constitute the egg apparatus, which consists of the egg cell flanked by the two synergids. Three of the four nuclei at the chalazal end of the gametophyte give rise to the so-called antipodal cells. Such is the structure of a "normal" eight-nucleate female gametophyte. It is known as the *Polygonum* type of monosporic female gametophyte. Another variant of the monosporic female gametophyte is the *Oenothera* type, which is characteristic of the family Onagraceae. Usually the micropylar rather than the chalazal megaspore of the tetrad is the functional one, and only two (rather than three) successive nuclear divisions result in a distinctive type of gametophyte consisting of the three-celled egg apparatus and a single polar nucleus. Because there are only two nuclear divisions, the second polar nucleus and the antipodal cells are not formed. The *Oenothera* type evidently derived from the *Polygonum* type as a result of the abbreviation of ontogeny.

In the bisporic type of gametophyte development, one of the haploid binucleate dyad cells formed after the first meiotic division gives rise to the gametophyte. This is connected with the complete failure of wall formation after the second meiotic division and, therefore, both the nuclei of the functioning dyad cell participate in the formation of the female gametophyte. The development of the bisporic type occurs in a shorter way, and for the formation of the eight-nucleate gametophyte only two successive mitotic divisions are required. Thus, the bisporic type is more advanced than the monosporic one. It is characteristic for the Podostemaceae, Viscaceae, Limnocharitaceae, Alismataceae, *Hyacinthoides (Endymion)*, *Allium*, Amaryllidaceae, *Convallaria*, and some other taxa.

In the most advanced and specialized tetrasporic type of the gametophyte development, the entire tetrad of megaspores, not separated by cell walls, takes part in forming the mature gametophyte. The mother cell of the megaspores themselves is transformed in this case into an extremely distinctive four-nucleate coenomegaspore. After meiosis, the four megaspore nuclei exhibit three main types of arrangement in the coenomegaspore. Each type of arrangement of the nuclei, except the so-called *Adoxa* type, gives rise to different (about ten) types of the mature gametophyte structure. The tetrasporic types evolved in very different lines of angiosperm evolution and characterize, for example, *Piper, Peperomia, Caulophyllum, Plumbago, Plumbagella, Tamarix, Ulmus, Penaea, Adoxa, Drusa, Chrysanthemum, Anthemis, Gunnera, Medeola, Lilium, Fritillaria, Maianthemum*, and many others.

It is generally agreed that the most common monosporic

eight-nucleate *Polygonum* type of female gametophyte represents the basic and primitive form in the flowering plants. But even this primitive type is very different from the typical female gametophyte of gymnosperms. Only two gymnospermous genera, *Welwitschia* and *Gnetum,* have female gametophytes which resemble in some respects those of the flowering plants. This resemblance consists not only in the absence of archegonia but also in the tendency towards the decrease in the number of nuclei. Both these types of female gametophytes remind us of an early stage of development of the female gametophyte of archegoniate gymnosperms, possessing a peripheral layer of free nuclei arranged around a large central vacuole. It is therefore quite possible that the nonarchegoniate angiosperm gametophyte originated by way of progressive acceleration of the gametogenesis and retardation of all other developmental processes.

Similar ideas were expressed long ago and can be found, for example, in the works of Strasburger (1900) and Coulter (1909, 1914) and some even earlier authors. In his almost forgotten book, "The evolution of sex in plants," Coulter says that the "complete elimination of the archegonium, begun in gymnosperms, is a feature of all angiosperms" (1914:73). The archegonium develops at a progressively earlier stages of development, "until finally eggs are matured before there is any tissue to develop the sterile jacket." In other words, says Coulter, "the archegonium is reduced to its essential sexual structure, the egg, which means that the distinguishing feature of an archegonium, the jacket, has disappeared" (p. 73). The only correction we should make is that, strictly speaking, the angiosperm egg is not the former egg of the archegonium but one of the cells of the initial developmental phase of the gametophyte. But this was clear already to Hoffmeister (1858) and Strasburger (1900) and has been convincingly shown lately by Gerasimova-Navashina (1958, 1971). Various versions of the "archegonial disappearance theory," as it was called by Battaglia (1951), have been accepted by Maheshwari (1948, 1950), Battaglia, and some others.

All these facts and considerations lead to the conclusion that the nonarchegoniate female gametophyte originated by neoteny (Romanov 1944, Takhtajan 1948, 1954, and later publications, summarized in 1976, Zimmermann 1959). According to the neotenic version of the "archegonial disappearance theory," the nonarchegoniate female gametophyte emerged through the delay in the development of the ancestral gametophyte at its initial free nuclear stage and subsequent formation of cellular structures of an entirely new type. It was not only a hereditary fixation of the early stage but a fundamental change of the entire course of development of the gametophyte. The main change here consisted in abrupt terminal abbreviation, as a result of which all the late stages of development—including the formation of the archegonium—were omitted. Thus the female gametophyte of the flowering plants does not attain in its development that stage of cellular differentiation at which the formation of archegonia begins. Moreover, the egg is differentiated no later than the third division of the megaspore nucleus, i.e., at such an early stage that the initiation of archegonia is not yet possible. Thus, it is natural that the archegonium is eliminated, leaving no trace.

At first, the basal abbreviation did not take part in the origin of the female gametophyte of flowering plants. But having originated, the female gametophyte underwent further reduction in its development also by this means. The basal abbreviation is realized in quite an original manner—it takes place owing to the inclusion of the products of megasporogenesis in the formation of the female gametophyte, which is observed in bisporic and particularly in tetrasporic types (figure 21).

The simplification of the female gametophyte of flowering plants and sharply expressed acceleration of its development are

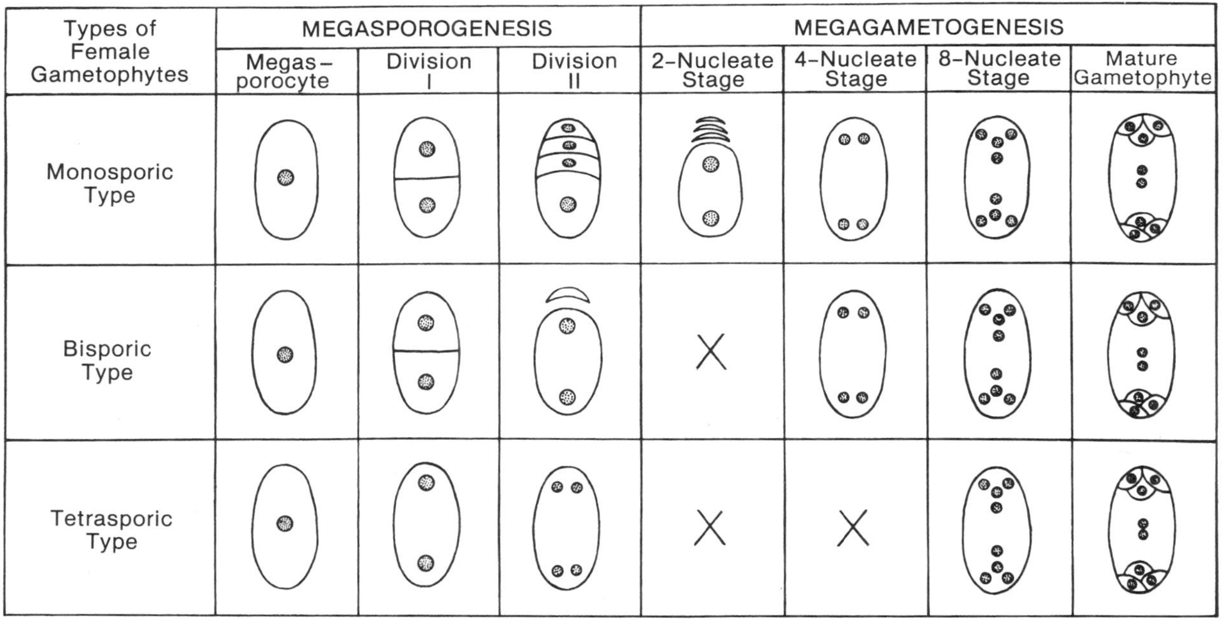

Figure 21. Diagrammatic comparison of the main types of megasporogenesis and megagametogenesis in flowering plants (from A. Takhtajan 1959).

the consequence of a quick maturation of the comparatively small ovules. Again, the early maturation of the latter and their small size are connected with the rapid development of the angiosperm flower, and with the presence of a closed carpel. Here is a definite chain of interconnected evolutionary changes, at the base of which lies the neotenous transformation of the strobilus of gymnospermous ancestors into the angiosperm flower.

6.3. Penetration of the Pollen Tube, Fertilization, and Triple Fusion

The tube cell of the mature male gametophyte of the germinating pollen grain gives rise to the pollen tube. In a vast majority of cases, the pollen tube penetrates into the megasporangium through a micropyle (porogamy), which is a primitive condition. More rarely, it enters by some other route (aporogamy). Aporogamy, which is a derived condition, is divided into chalazogamy and mesogamy. In chalazogamous plants, the tip of the pollen tube enters through the chalazal end of the ovule, rises upward, and continues its growth along the surface of the female gametophyte before reaching the egg apparatus. Chalazogamy is observed in the Casuarinaceae, Betulaceae, Juglandaceae, and some other "amentiferous" taxa. In rather rare mesogamous plants, such as *Circaeaster, Alchemilla, Cucurbita,* and others, the pollen tube enters laterally between micropyle and chalaza.

After entering the female gametophyte and reaching the egg apparatus, the pollen tube discharges its contents, including the two male gametes and the tube cell nucleus. One of the two male gametes fuses with the egg cell (syngamy), and the other, with two polar nuclei (or with the secondary nucleus) of the central cell of female gametophyte (triple fusion). As a result of syngamy, a diploid zygote is formed; and as a result of triple fusion, triploid primary nucleus of the endosperm. This triple fusion, resulting in a triploid endosperm, is a characteristic feature of magnoliophytes, sharply distinguishing them from all other plants. "The fertilization" of the central cell, i.e., fusion of the second male gamete with two free or already fused polar nuclei is in fact not true fertilization, as it does not give rise to an embryo, but an evolutionary new and very special structure, the endosperm, a food storage tissue whose nutrients are absorbed by the embryo developing from the fertilized egg. But there are some derivative types. In the *Oenothera* type of female gametophyte, there is only a single polar nucleus and the endopserm is diploid rather than triploid. In tetrasporic female gametophytes there occur tetraploid (*Penaea* and *Plumbago* types) and even octoploid (*Peperomia* type) secondary nucleus of the central cell. But in an overwhelming majority of flowering plants, the central cell has only two nuclei and the endosperm is therefore triploid; although in the Podostemaceae and Orchidaceae, endosperm fails to develop or degenerates early in ontogeny.

It was suggested by Strasburger (1900) that the triple fusion serves as "vegetative fertilization" for stimulation of rapid development of the endopserm. The union of the male gamete with the polar nuclei undoubtedly stimulates a rapid development of the nutritive tissue. Triple fusion originated as a result of neotenous simplification of the female gametophyte, which is usually almost completely devoid of reserve materials and is an adaptation for quick compensation of nutritive materials. Consequently, the origin of endosperm is correlated with an extreme simplification of the female gametophyte and is a singular compensatory device (see Brink and Cooper 1940, 1947). Thus, the origin of "double fertilization" was a consequence of the extreme simplification and miniaturization of the female gametophyte of flowering plants. It permitted attaining maxium economy of material for constructing the highly organized female gametophyte and an intense acceleration of its formation.

References

Battaglia E. 1951. The male and female gametophytes of angiosperms —an interpretation. Phytomorphology 1:87–116.

Brewbaker J. L. 1967. The distribution and phylogenetic significance of binucleate and trinucleate pollen grains in the angiosperms. Amer. J. Bot. 54:1069–1083.

Brink R. A. and D. C. Cooper. 1940. Double fertilization and development of the seed in angiosperms. Bot. Gaz. 102:1–25.

Brink R. A. and D. C. Cooper. 1947. The endosperm in seed development. Bot. Rev. 13:423–541.

Coulter J. M. 1909. Evolutionary tendencies among gymnosperms. Bot. Gaz. 48(2):81–97.

Coulter J. M. 1914. The evolution of sex in plants. Chicago.

Coulter J. M. and C. J. Chamberlain. 1912. Morphology of Angiosperms. New York.

Cronquist A. 1988. The evolution and classification of flowering plants. 2d ed. New York.

Eames A. J. 1961. Morphology of the angiosperms. New York.

Gerasimova-Navashina E. N. 1958. On the gametophyte and on salient features of development and functioning of reproducing elements in angiospermous plants. Probl. Bot. (Leningrad) 3:125–167. (In Russian.)

Gerasimova-Navashina E. N. 1971. Double fertilization in angiosperms and some of its theoretical aspects. In: V. P. Zosimovich (ed.), Problems in embryology. Kiev, pp. 113–152. (In Russian).

Hoffmeister W. 1858. Neuere Beobachtungen über die Embryobildung der Phanerogamen. Jahrb. f. wiss. Bot. 1:82–190.

Johri B. M. 1963. Female gametophyte. In P. Maheshware, ed., Recent advances in the embryology of angiosperms, pp. 69–103. Delhi.

Maheshwari P. 1948. The angiosperm embryo sac. Bot. Rev. 14:1–56.

Maheshwari P. 1950. An introduction to the embryology of angiosperms. New York.

Romanov I. D. 1944. Evolution of the embryo sac of flowering plants.

Unpublished doctoral dissertation. Tashkent University. (In Russian.)

Romanov I. D. 1971. Types of development of the embryo sac of angiospermous plants. In V. P. Zosimovich, ed., Problems in Embryology, pp. 72–113. Kiev. (In Russian.)

Rudenko R. E. 1959. The significance of the male gametophyte for the taxonomy of Angiospermae. Biol. Zhurn 44:1467–1475. (In Russian.)

Schnarf K. 1939. Variation in Bau des Pollenkornes der Angiospermen. Tabul. Biol. 17:72–89.

Stebbins G. L. 1957. Self-fertilization and population variability in the higher plants. Amer. Natur. 91:337–354.

Stebbins G. L. 1974. Flowering plants. Evolution above the species level. Cambridge, Mass.

Strasburger E. 1900. Einige Bemerkungen zur Frage nach der "Doppelten Befruchtung" bei den Angiospermen. Bot. Zeitung 58:293–316.

Takhtajan A. 1948. Morphological evolution of the angiosperms. Moscow. (In Russian.)

Takhtajan A. 1954. Quelques problèmes de la morphologie évolutive des angiospermes. Voprosy Botan. 2:763–793. (In Russian and French.)

Takhtajan A. 1959. Die Evolution der Angiospermen. Jena.

Takhtajan A. 1976. Neoteny and the origin of flowering plants. In C. B. Beck, ed., Origin and early evolution of angiosperms, pp. 207–219. New York and London.

Willemse M. T. M. and J. L. van Went. 1984. The female gametophyte. In B. M. Johri ed., Embryology of Angiosperms, pp. 159–196. Berlin.

Zimmermann W. 1959. Die Phylogenie der Pflanzen. 2 Aufl. Stuttgart.

7

Evolution of Fruits

The fruit is formed from the flower as a result of structural and functional changes of the gynoecium and frequently, also, of some extracarpellary organs. In the most primitive fruits, like those of Magnoliaceae, only the carpels and the floral axis participate in the formation of the fruit. But during evolution—in connection with the growing specialization of flowers, including the emergence of the inferior ovary and various devices for protection and dispersal of seeds—other parts of the flower participate more and more in the formation of fruits. Therefore, the fruit should be considered as the product of the flower and not of the gynoecium only, though its basic parts are formed by the carpels. Some authors define a fruit as a "mature flower."

The fruit structure preserves mainly the features of those parts of the flower which have given rise to it. Though the outlines of the carpels are often quite distinct in the anatomical structure of the fruit and its conducting system repeats the vascularization of the gynoecium, the mature fruit differs from the corresponding floral parts as greatly as the seed-coat from the integuments.

The fruit wall (pericarp) may be parenchymatous and succulent or it may be dry. On this basis, fruits are usually divided into fleshy and dry, subdivided in their turn into drupes, berries, capsules, nuts, etc. Such a division, however, is purely artificial and based on similar adaptive features appearing in various lines of evolution. It does not reflect their evolutionary interrelationships to any extent. The evolutionary classification should be based not on the degree of solidity of the pericarp and the modes of its opening or some other ecologically defined traits but on origin of each fruit type. A whole series of attempts to create an evolutionary classification of the fruit types was made (Harvey-Gibson 1909; Gobi 1921; Gusuleac 1939a, 1939b; Winkler 1939, 1940; Egler 1943; Kaden 1947, 1961; Takhtajan 1948, 1959; Baumann-Bodenheim 1954; Levina 1961, 1987). But the classification of fruits is not yet worked out sufficiently to cover their diversity and to show the evolutionary interrelationships of all their types. However, the main types of fruits and their general evolutionary trends can be taken now as established.

If the fruit is formed from an apocarpous gynoecium, the individual carpels remain in most cases distinct and at maturity each of them forms a separate fruitlet (carpidium). Such fruits are termed apocarpous. If, however, the fruit is formed from the syncarpous gynoecium or if during development the ab initio free carpels coalesce, the fruits are termed syncarpous.

7.1. Apocarpous Fruits

The most primitive and basic fruit type is a fruit consisting of many-seeded distinct follicles (Hallier 1902, 1912; Harvey-Gibson 1909; Bessey 1915; Gobi 1921; and many others). Such a fruit, developing from a multicarpellate apocarpous gynoecium, was called multifollicle by Gobi. The most primitive form of multifollicle consists of numerous, individual fruitlets (carpidia), spirally arranged on a floral axis (e.g., *Magnolia, Trollius, Caltha, Paeonia*). Each follicle is many-seeded or, due to reduction, with two or even one seed, and commonly opens from the apex towards the base ventrally (i.e., along the suture) or more rarely dorsally (e.g., *Manglietia, Magnolia, Alcimandra, Michelia*).* The spiral multifollicle evolved into a cyclic multifollicle in which the fruitlets are arranged in whorls around a shortened axis (e.g., *Illicium, Spiraea, Crassula*). The multifollicle gave rise to unifollicle by reduction in the number of carpels (e.g., *Degeneria, Consolida, Cercidiphyllum*).

In some magnoliids and ranunculids, the follicles are somewhat berrylike, more or less fleshy, and usually indehiscent or succulent. The succulent multifollicle characterizes most of the Annonaceae, *Austrobaileya, Schisandra, Decaisnea, Sargentodoxa, Hydrastis,* and others. In the genus *Decaisnea* of Lardizabalaceae, each fruitlet opens ventrally thereby indicating the origin from a typical follicle. Only in a few genera are there found the succulent unifollicles (*Degeneria, Actaea*). The succulent follicle probably appeared very early in angiosperm evolution.

Another direction of follicular specialization led to the formation of the legume—the type of follicle characteristic of the Fabales. The legume differs from the typical follicle only by its dehiscence—it opens simultaneously along the suture of carpel margins and along the dorsal vein. The advantage of such a method of dehiscence lies in the fact that the valves of a dried legume usually dehisce explosively; the valves instantaneously contort and the seeds are expelled (autochory).

In certain groups of Fabales, the usual type of dehiscent legume undergoes various specializations connected with the methods of seed dispersal. Thus, in certain Caesalpinioideae, such as *Gymnocladus, Ceratonia,* and *Tamarindus,* indehiscent succulent legumes are found. The succulence is an adaptation for endozoochory.

---

*In *Kmeria,* follicles open ventrally and partly also dorsally. In *Talauma,* woody fruitlets, which are already more or less coalescent (at least at the base), dehisce transversally (circumscissile dehiscence).

In representatives of the tribe Detarieae (Caesalpinoideae), and in some members of the tribe Swartzieae (Faboideae), a very modified indehiscent legume is drupaceous. The jointed legumes of certain Mimosoideae and Faboideae, such as *Entada,* many species of *Mimosa, Ewersmannia subspinosa, Hedysarum, Coronilla,* and *Ornithopus,* are no less peculiar. In a majority of the members of the tribe Hedysareae, the legume (called a loment) is strongly contracted between the seeds. Quite often, the legumes are transformed into a single-seeded indehiscent nutlike fruit, as in the genus *Onobrychis.* Winged nutlike fruits adapted to dispersal by wind are also met with (especially in the tribe Dalbergieae). The utricularly inflated legumes of *Smirnovia* and *Colutea* also are anemochorous.

Another direction of the evolution of the follicular fruits is connected with a reduction in the number of seeds in each fruitlet to one, with the consequent loss of capacity for active dehiscence. Hence the multifollicle gave rise to multiachenium which consists of achenae,* i.e., single-seeded and indehiscent fruitlets with a leathery or more or less woody pericarp. The multiachenium characterizes *Anemone, Clematis, Adonis, Ranunculus,* and some other Ranunculaceae, *Potentilla, Geum,* and some other Rosaceae-Rosoideae, as well as Alismataceae, *Soridium* (Triuridaceae), etc. The inner and the intermediate layers of the pericarp of the achenium consist of more or less lignified —often sclerenchymatous—cells, whereas the outer one is of the thin-walled nonlignified parenchyma.

---

*I agree with Fahn, who defines an achenium or akenium (achene or akena) as "a single-seeded fruit formed by one carpel (*Ranunculus*)" (1974:508). But many authors apply this term also to the morphologically quite different fruits of Asteraceae, which develops from bicarpellate inferior ovary surrounded by other floral tissues and, therefore, greatly differ from a true achenium. The term multinucula (multinutlet), used by some authors including myself, is also less appropriate than multiachenium (nucula is a diminutive form of nux, which is a syncarpous fruit).

There are some intermediate forms between multifollicles and multiachenia, as in *Liriodendron* and *Platanus*.

A special type of multiachenium is the fruit of *Fragaria* and *Duchesnea* where numerous fruitlets—the achenia—are situated on the fleshy and succulent receptacle that expands after flowering. Such a fruit is named fragum. Individual fruitlets of the fragum have the same structure as the other achenia. The fleshy receptacle of the fragum is an endozoochorous adaptation.

The fruit of the genus *Nelumbo,* where each of the achenia is submerged in the extending receptacle, is a very special type of multiachenium. Such a fruit is called a submerged multiachenium.

In cases where the individual fruitlets of the multiachenium are situated on the inner surface of the saucer-shaped or more or less tubular fleshy "hypanthium" as in Monimiaceae, Calycanthaceae, and *Rosa,* the fruit is called cynarrhodium (from gr. *cynarrhodon*—dog rose). The cynarrhodium as well as the fragum differs from the usual multiachenium only in the arrangement of carpels, not in their structure.

Lastly, the drupe originated from the follicle as a result of succulence of the mesocarp, lignification of the endocarp, and decrease of the number of seeds. In certain cases, the drupe could have originated from the achenium. The inner part of the pericarp of the drupe or its endocarp is a hard woody putamen which is followed by a more or less fleshy or even succulent mesocarp, surrounded by a coriaceous exocarp. The drupe, unlike the follicle, is mostly single-seeded and has lost the capacity for active dehiscence except in "dry drupe" of *Amygdalus* and some cultivated varieties of peaches. The fruit of the subfamily Prunoideae of the family Rosaceae is a typical drupe. In most genera of this family, the fruit is unidrupaceous, but in the archaic genus *Osmaronia,* it is multidrupaceous. The multidrupaceous fruit characterizes *Amborella,* many Menispermaceae, the genera *Rhodotypos, Keria, Rubus,* and some other taxa.

7.2. Dry Syncarpous Fruits

The multifollicle gave rise to a vast and very diversified group of syncarpous capsules as a result of ontogenetic or congenital fusion of carpels. Only in a few cases, as in the genus *Pachylarnax* (Magnoliaceae) or in some Winteraceae, syncarpous capsules evolved from spiral multifollicles. Mostly, they originated from cyclic multifollicles like those of *Illicium.*

An intermediate form between the multifollicle and typical syncarpous fruit is the syncarpous multifollicle, which at the time of ripening, dehisces through the sutural areas of the upper free parts of the fruitlets. The fruits of some species of *Zygogynum* (Winteraceae), of *Trochodendron* and *Tetracentron, Nigella,* some species of *Spiraea,* etc., are examples of the syncarpous multifollicle.

It is only one step from the syncarpous multifollicle to the capsule, which is the most primitive type of dry syncarpous fruit. The true capsule differs from the syncarpous multifollicle by a more complete coalescence of the individual fruitlets and a more specialized dehiscence of the fruit.

There are three basic types of the capsule—septate with axile placentation (syncarpous sensu stricto), unilocular with parietal placentation (paracarpous), and unilocular with free-central placentation (lysicarpous).

The septate syncarpous capsule is mostly tri-quinquelocular, more rarely bi- or multilocular. The dehiscence of the mature capsule occurs in various ways (Kaden 1962). Dehiscence may occur longitudinally along the line of union of adjacent carpels, or along the dorsal vein of each carpel, or along the suture and the back simultaneously. Separation along the suture is the most primitive method of dehiscence. The dehiscence through septa is called septicidal dehiscence (e.g., *Veratrum*), and dehiscence through the backs of locules is called loculicidal dehiscence (e.g., *Iris*). Loculicidal dehiscence is a more advanced type than the septicidal one. Even more specialized a type is the circumscissile dehiscence by the formation of a transverse ringlike crack in the pericarp which results in the formation of a lid, e.g., in *Portulaca, Anagallis, Hyoscyamus,* or *Plantago.* This type of capsule is called lid capsule or pyxis (pyxidium). In some cases, as in *Anthirinum* and *Linaria,* the capsule dehisces through pores which develop in the pericarp (poricidal dehiscence).

In some cases, as in a majority of Euphorbiaceae and in many Malvaceae, the mature septate capsule divides along the septa into separate one-seeded halves (mericarps), which dehisce along the dorsal slit. This kind of a capsule is called regma or schizocarp.

The next step in the evolutionary specialization of capsules was the origin of those fruits falling apart into separate indehiscent parts while ripening. In some cases, the number of these parts corresponds to the number of locules, while in others, by forming a false septum, each locule (each carpel) is divided in its turn into two halves as in Boraginaceae and Lamiaceae. In the latter, each unit of dispersal is a semimericarp or eremus. To denote the fruits of this type, Kaden (1961) used the term coenobium. In Aceraceae, the mericarps of the bilocular fruit are winged and so the fruit is called bisamara.

The fruit type called samara is derived from the pseudomonomerous gynoecium, where a thin winglike margin is formed along the borders of pericarp during fruit development (e.g., *Eucommia, Ulmus, Hemiptelea, Pteroceltis*).

Somewhat less diversified are the syncarpous dry fruits evolved from the inferior ovary. The most common type is the inferior septate capsule.

Very specialized types of inferior septate syncarpous fruits are acorn (glans) and nut (nux). These indehiscent single-seeded fruits are formed from a septate gynoecium by an underdevelopment of locules and ovules. The acorn has a coriaceous pericarp and is surrounded by a cuplike organ termed a cupule. An example is the acorn of *Quercus*. In the nut, the pericarp is more or less woody and strongly lignified. *Corylus* serves as an example.

A very distinctive fruit type is the so-called balausta typical in the genus *Punica*. The pericarp of the balausta is dry and opens by irregular cracks, while the radially elongated cells of the outer epidermis of the seed coat form a fleshy layer (pulpa).

Many of paracarpous or nonseptate capsules with parietal placentation originated from the septate syncarpous fruits with axile placentation. But in some archaic groups, paracarpous fruit originated directly from the apocarpous one. Good examples are: *Mondora* and *Isolona* (Annonaceae), Canellaceae, and *Takhtajania* (Winteraceae). Thus the paracarpous type is of dual origin.

In typical cases the paracarpous capsule is unilocular and the placentation is strictly parietal. Dehiscence of paracarpous capsules usually occurs along the midribs of the carpels (dorsicidal dehiscence), as in *Viola*, Tamaricaceae, Salicaceae, Gesneriaceae, and many others. Sometimes the paracarpous capsule dehisces, however, by means of small pores formed in the pericarp, as in *Papaver*. Such a type of paracarpous capsule with poricidal dehiscence was called papaverella.

In many groups, the typical paracarpous capsule gives rise to capsules with false septa. These false septa are the result of expansion and protrusion of parietal placentae into the cavity of the ovary and their more or less coalescence in its center. A good example of such secondarily septate capsules are the inferior capsules of the Campanulaceae.

In some groups, the paracarpous capsule give rise to paracarpous nutlike fruits. There are both the capsules (e.g., *Dicentra*,

Corydalis) and nutlike fruits (e.g., *Fumaria*) within the family Fumariaceae. The paracarpous fruits of *Carex* and some other Cyperaceae also are nutlike. They are called utricle.

The paracarpous capsule gave rise to the siliqua, the bicarpellate paracarpous fruit characterizing a majority of the Brassicaceae. The placentae of siliqua forms a thick rib, termed a replum, around the fruit. From the placentae, two membranes grow inwards where they fuse to form a false septum that divides the locule into two chambers. The siliqua commonly dehisces from below upwards, leaving the seeds attached to the replum. The replum itself remains as a frame around the septum. A more primitive type of siliqua with a well-developed replum but without a false septum is found in the subfamily Cleomoideae of the family Capparaceae, e.g., in *Cleome* and *Polanisia*. In certain Brassicaceae are found various modifications of the siliqua, including the constrictions formed between the seeds or transverse walls, forming one-seeded segments that separate when ripe into joints which may be indehiscent, as for example in *Cakile*.

The caryopsis of the grasses is a special type of the paracarpous fruit. It is an indehiscent monospermous fruit in which the thin pericarp is so closely attached to the seed coat that it seems to have fused with it. The caryopsis evolved from the paracarpous capsule.

One of the most specialized paracarpous fruits is cypsela of the Asteraceae, a single-seeded bicarpellate indehiscent fruit developing from an inferior ovary.

In certain magnoliopsids the septate capsule gave rise to the lysicarpous capsule (unilocular with free-central placentation). The transitional forms between the syncarpous capsule and the lysicarpous are found in certain representatives of the subfamily Silenoideae of the family Caryophyllaceae while lysicarpous capsules are typically found in the families Portulacaceae and Primulaceae. A majority of lysicarpous capsules dehisce by forming teeth or valves in the upper part of the fruit. In the genera *Portulaca* and *Anagallis*, the fruit dehiscence is circumscissile. The Plumbaginaceae are characterized by a lysicarpous nutlike fruit.

7.3. *Fleshy Syncarpous Fruits*

In many lines of angiosperm evolution various types of syncarpous dry fruits gave rise to endozoochorous fleshy fruits. In certain groups of both magnoliopsids and liliopsids, the syncarpous drupes result from the syncarpous capsule. Unlike the true drupe of *Prunus*, which originates from one carpel, the syncarpous drupe appears from several carpels. Therefore the stone (or the stones) of the syncarpous drupe should be better called pyrenes (as against the putamen of the apocarpous drupe). But —in structure—it is often very much like the drupe of *Prunus*, which can be explained by convergence. The most primitive types of syncarpous drupes are polypyrenous and the advanced are monopyrenous. The fruits of Aquifoliaceae (four to eight stones) and Rhamnaceae (two to four stones), for example, are polypyrenous drupes. The drupes with two or three (to ten) seeds gave rise to the monospermous ones in some palms. Very rarely a schizocarpous or split drupe is found, e.g., in *Cneorum, Kirkia,* some Colletieae (Rhamnaceae), and some Myoporaceae.

Syncarpous berries are another type of syncarpous fleshy fruits. Unlike the drupes, the berry does not have a stone. The anatomy of the berry often shows vestigial structures, revealing its origin from the capsule (see particularly Zazhurilo 1936). Typical syncarpous berries are mostly polyspermous and characterized by entirely fleshy or succulent pericarp at the ripe stage. Fruits of Ebenaceae, Sapotaceae, some Vitaceae, *Lycopersicon* and other Solanaceae, *Convallaria, Ruscus, Dracaena, Paris, Phoenix,* and some other Arecaceae, and many other flowering plants belong to the syncarpous berry type. In some lines of evolution, the typical unilocular paracarpous capsule gives rise to the paracarpous berries. They are found in many Flacourtiaceae, some Violaceae, *Passiflora, Capparis,* etc.

The hesperidium, characterizing the subfamily Citroideae of the family Rutaceae is a berrylike syncarpous fruit. The exocarp of the hesperidia (flavedo) consists of epidermis with very small, thick-walled cells and compact subepidermal parenchyma with essential oil glands; the mesocarp (albedo) consists of parenchyma with large intercellular spaces and vascular network; and the endocarp consists of inner epidermis and compact parenchyma. The inner epidermis and subepidermal layers of endocarp produce the large multicellular spindle-shaped stalked juice sacs which fill the locules when the fruit ripens.

Besides the above-listed superior syncarpous fleshy fruits a considerable number of types develops from the flowers with an inferior ovary as well. Thus the inferior drupe is represented in *Davidia,* Cornaceae, *Curtisia, Mastixia, Alangium, Aralidium, Toricellia, Helwingia,* many Araliaceae, etc. The inferior berry (typical bacca) characterizes the subfamily Vaccinioideae of the family Ericaceae, *Griselinia, Aucuba, Garrya,* etc.

The pome characteristic of the subfamily Maloideae of the family Rosaceae is a relatively primitive—and, at the same time, very distinctive—inferior syncarpous fleshy fruit. The pome originated from a syncarpous multifollicle as a result of its overgrowth with the floral tube. Both the floral tube and the outer layers of the carpellary tissue turn fleshy on ripening, while the inner layers of the carpellary tissue become cartilaginous and at times hard and consist of the fibrous sclereids (MacDaniels 1940). The conversion of the syncarpous multifollicle into the pome was connected with the transition to endozoochory. In a typical pome of *Malus* type, the endocarp is cartilaginous, while the outer fleshy layers of the carpel unite with the tissue of the floral tube. In other cases, as—for example—in the fruits of *Cotoneaster* and *Crataegus,* each carpel forms a solid stone of pyrena type from the lignified sclerenchyma (pomum pyrenatum).

A special type of the inferior paracarpous berry is the pepo, a polyspermous fruit with a juicy endocarp, a fleshy mesocarp, and a more or less hard exocarp. The pepo characterizes many Cucurbitaceae. The paracarpous drupe is also known. Fleshy fruits originated also from the caryopsis (berrylike fruits of certain Bambuseae).

From the araliaceous syncarpous drupe, there originated a very distinct dry fruit called cremocarp. It characterizes most of the Apiaceae family and some representatives of the Araliaceae, like the genera *Stilbocarpa* and *Myodocarpus.* It consists of two indehiscent one-seeded mericarps which usually separate in the mature stage in the plane of union of the two carpels. Before falling, the two mericarps usually hang for a time on a commonly forked stalk, the carpophore, which develops from the carpellary tissues.

References

Baumann-Bodenheim M. G. 1954. Prinzipien eines Fruchtsystems der Angiopsermen. Ber Schweiz. Bot. Ges., 64:94–112.

Bessey C. E. 1915. The phylogenetic taxonomy of flowering plants. Ann. Missouri Bot. Gard. 2:109–164.

Bischoff G. W. 1834. Lehrbuch der Botanik. I, 1. Allgemeine Botanik. Stuttgart.

Egler F. E. 1943. The fructus and the fruit. Chron. Bot. 7: 391–395.

Fahn A. 1977. Plant anatomy. 2d ed. Oxford.

Gobi Ch. 1921. The genetical classification of fruits. Zapiski Labor. Semenoved. IV, 4:5–30. (In Russian with French summary.)

Gusuleac M. 1939a. Der genetisch Standpunkt in Taxonomie der Früchte. Bull. Fac. Sti. Cernäaut 12:206–219.

Gusuleac M. 1939b. Zur Präzisierung der Nomenklatur der Früchte und der Prinzipien eines natürlichen Früchtsystems. Bull. Fac. Sti. Cernäaut 12:337–355.

Hallier H. 1902. Beiträge zur Morphogenie der Sporophylle und des Trophophylle in Beziehung zur Phylogenie der Kormophyten. Jahrb. Hamburg. Wiss. Anst. XIX, 3. Beiheft, 1–110.

Hallier H. 1912. L'origine et le système phylètique des Angiosperms exposés à l'aide de leur arbre généalogique. Arch. Néerl. Sci. Exact. Nat. 3d ser., 1:146–234.

Harvey-Gibson R. I. 1909. A classification of fruits on a physiological basis. Trans. Liverpool Bot. Soc. 1:1–5.

Juhnke G. and H. Winkler. 1938. Der Balg als Grundelement des Angiospermengynaeceums. Beitr. Biol. Pfl. 25:29–324.

Kaden N. N. 1947. Genetical classification of fruits. Vestn. Mosk. Univ. 12:31–42. (In Russian.)

Kaden N. N. 1961. On some main problems of classification, typology, and nomenclature of fruits. Bot. Zhurn. (Leningrad). 46(4):496–504. (In Russian.)

Kaden N. N. 1962. Types of longitudinal dehiscence of fruits. Bot. Zhurn. (Leningrad). 69(4):495–505. (In Russian.)

Levina R. E. 1961. On the classification and nomenclature of fruits. Bot. Zhurn. (Leningrad). 46(4):488–495. (In Russian.)

Levina R. E. 1987. Morphology and ecology of fruits. Leningrad. (In Russian.)

MacDaniels L. H. 1940. The morphology of the apple and other pome fruits. Ithaca, N. Y. (Cornell) Agr. Exp. Sta. Mem. 230.

Takhtajan A. 1948. Morphological evolution of the angiosperms. Moscow. (In Russian.)

Takhtajan A. 1959. Die Evolution der Angiospermen. Jena.

Winkler H. 1939. Versuch eines natürlichen Systems der Früchte. Beitr. Biol. Pfl. 26:201–220.

Winkler H. 1940. Zur Einigung und Weiterführung in der Frage des Fruchtsystems. Beitr. Biol. Pfl. 27:92–130.

Zazhurilo K. K. 1931. On the classification of ornitochorous fruits and seeds. Zhurn. Russk. Bot. Obshch. 16:169–189. (In Russian.)

Zazhurilo K. K. 1936. Traces of evolution of fruits in their anatomical structure. Trudy Voronezhsk. Univ. 9(1):5–26. (In Russian.)

8

Evolution of the Seed

Following triple fusion, both the zygote and the primary nucleus of the endosperm develop further. The development of the seed begins with the divisions of the primary endosperm nucleus, followed by the emergence of embryo. The mature seed consists of an embryo, a more or less developed endosperm (which is completely absent in some taxa), and the seed coat. In addition, a special kind of storage tissue—the perisperm—is formed from the remnants of the nucellar tissue in some taxa.

The seeds of archaic flowering plants are of medium size, 5 to 10mm long (Corner 1976). Both small and large seeds are derived.

Evolution of the flowering plants was accompanied by an increase in the size of the embryo and a corresponding reduction, and even disappearance of the endosperm and other storage tissues. In more archaic flowering plants, seeds are characterized by a minute and undifferentiated embryo and an abundant endosperm (Pritzel 1898; Hallier 1912; Martin 1946; Eames 1961; Grushvitski 1961; and many others). In the advanced groups, on the contrary, the embryo is large and well differentiated, and the endosperm is more or less reduced or even wanting. Here, we observe something analogous to what happens in the animal world, where the embryo in the mother's body attains greater development in the higher forms (Nägeli 1884; Hallier 1902, 1912).

The reduction of endosperm is functionally correlated with an increase of the size of the embryo. The presence of a larger embryo renders it possible to deposit the reserve substances in the embryo itself in its cotyledons. This is quite a considerable advance, as the food deposited in the cotyledons is more accessible to the expanding embryo.

8.1. Emergence of the Monocotyledonous Embryo

The embryo in most flowering plants consists of a radicle, hypocotyl (portion of the stem), cotyledons, and a plumule (the apical meristem). Above the hypocotyl are situated two, one, or rarely (as in *Degeneria* and *Idiospermum*) several cotyledons. Only in very reduced and simplified embryos of a number of parasitic plants are there no traces of cotyledons.

On the basis of the structure of the embryo, the flowering plants are divided into two classes—the Magnoliopsida (dicotyledons) and the Liliopsida (monocotyledons). In the first, the embryo usually has two cotyledons (rarely, several or one); in the second it has only one. It is widely accepted that the monocotyledonous embryo arose from the dicotyledonous embryo.

On the basis of a comparative study of the embryos of various dicotyledons and monocotyledons, Hegelmaier (1874, 1878) concluded that the monocotyledonous embryo resulted by suppression or loss of one or two cotyledons of the typical dicotyledonous plants. This so-called "Abort-Hypothese" later gathered wide support. It was developed by Henslow (1893, 1911), Winkler (1931), Metcalfe (1938), Yakovlev (1946), Eames (1961), and many others.

All the available data lead to the conclusion that the old

"underdevelopment hypothesis" of Hegelmaier is fully confirmed. We find its confirmation, above all, in the dicotyledonous taxa, where—in some species and, even, in genera—the monocotyledonous embryo is normally developed. Such are *Ficara*, some species of *Peperomia, Corydalis, Claytonia, Cyclamen, Bunium, Pinguicula*, some Gesneriaceae of the tribe Cyrtandreae (Hegelmaier 1878; Metcalfe 1938; Haccius 1954; Crété 1956; etc.).

The studies of the early stages of development of the embryo in many liliopsids also confirm the underdevelopment hypothesis. As the numerous works showed, the first phases of embryo development—the formation of two-celled proembryo—are similar in magnoliopsids and liliopsids. As has been particularly stressed by Yakovlev (1946), the initial or proembryonal stage is the most conservative stage of the embryonal development and is, as a rule, completely identical in the phylogenetically quite distant groups of flowering plants. But further development takes place in a different manner. In the magnoliopsids, two lateral cotyledons originated on both sides of the stem primordium, while, in the liliopsids, primordia of two cotyledons are noted only in the very beginning, and one of them was soon arrested in development and became indistinguishable. Numerous examples clearly show how embryo structure becomes asymmetrical and the only cotyledon becomes pseudoterminal. They show that the monocotyledonous embryo originated from the dicotyledonous embryo as a result of a sharply expressed early deviation (see also Haccius 1952 and Baude 1956).

It is very likely that the emergence of the monocotyledonous embryo is connected with the general simplification and reduction of the entire embryogenesis. This reduction in development is evidently related to the neotenous origin of the liliopsids.

8.2. Evolutionary Trends in Endosperm Formation

Three major types of the endosperm development are recognized—cellular, nuclear, and helobial.

In the cellular type, at least the first few divisions of the primary endosperm nucleus are accompanied by wall formation. The cellular pattern of development is found in many families of magnoliopsids (both archaic and advanced) and occurs only in three monocotyledonous families—Hydatellaceae, Araceae, and Lemnaceae.

In the helobial type, which is usually considered as somewhat intermediate between the cellular and nuclear types, the primary endosperm nucleus is always found at the chalazal end of the gametophyte; and, therefore, when it divides, two unequal cells or chambers are produced—a small chalazal cell and a much larger micropylar cell. The nucleus in the chalazal cell either does not divide further (a basic type, according to Swamy and Parameswaran 1963) or undergoes a usually restricted number of free nuclear divisions, whereas the larger micropylar cell undergoes numerous free nuclear divisions. Commonly, the cell-wall formation ultimately takes place in the micropylar chamber. The helobial type is common in liliopsids and is much less frequent in magnoliopsids.

In the nuclear type, the division of the primary endosperm nucleus is followed by a series of free nuclear divisions, resulting in the formation of a large multinucleate cell, which usually becomes cellular in a later phase of development. The nuclear type is widespread in both magnoliopsids and liliopsids.

The helobial type of endosperm development is probably apomorphic, derived either from nuclear or, more likely, from cellular type. But it is much more difficult to decide which of the two types—nuclear or cellular—is the more primitive. The main reason for this difficulty is that the formation of the endosperm is subject to reversal and that there are also many intermediates (Schnarf 1929, 1931; Brink and Cooper 1947). But, in spite of the reversibility of the types of endosperm development, the first flowering plants must have had either cellular or nuclear endosperm.

Schnarf (1929) and many others postulate that the nuclear pattern of endosperm development is more primitive than the cellular one. Sporne (1954), using the method of character correlations, found that nuclear endosperm shows highly significant positive correlations with woody habit, presence of secretory cells, stipules, free petals, number of stamens equal to—or more than—the number of perianth members, ovules with two integuments, and vascular bundles in the integuments. But later, Sporne (1967) came to the conclusion that there is a negative correlation between nuclear endosperm and two primitive characters—scalariform perforation of vessel members and apotracheal parenchyma. In 1980, Sporne also added negative correlations between nuclear endosperm and three other primitive characters—heterogenous rays, anatropous ovules, and albuminous seeds. "In view of the five negative correlations, 'Nuclear endosperm' is thus even more of an enigma now than it was twenty years ago, for no satisfactory explanation can be offered," states Sporne (Sporne 1967, p. 425).

Coulter and Chamberlain (1903), Schürhoff (1926), Rao (1938), Swamy and Ganapathy (1957), Wunderlich (1959), and others postulate that the evolutionary trend has been from the cellular to the nuclear type. Rao found a positive correlation between the type of embryo and the type of endosperm. In plants which have rapidly growing and early differentiating embryo, the endosperm is always nuclear. On the other hand, if the growth of the embryo is very slow and if the mature seed has only an undifferentiated embryo, the endosperm is either from the very beginning cellular or the wall formation begins very early. The rapid differentiation of embryo requires rapid consumption of endosperm and, therefore, the nuclear endosperm is more economical. The formation of cell walls in nuclear endosperm usually begins only when the embryo is already completely differentiated. But in cases of the slow growing or undifferentiated embryo, where the consumption of the endosperm takes place later, the cellular endosperm is formed. These correlations clearly show that the cellular endosperm is primitive. There are also other character correlations which result in the same conclusion.

Swamy and Ganapathy (1957) found that the cellular type is positively correlated with vessel members having scalariform perforations and apotracheal wood parenchyma, whereas the nuclear type shows a negative correlation with these primitive kinds of perforation and wood parenchyma.

Still another type of correlation was found by Wunderlich (1959), who studied the correlation of the endosperm type and the morphology of the ovule itself and came to the conclusion that cellular endosperm is correlated with a small nucellar cavity and the nuclear endosperm with a large cavity. She regards the presence of a small cavity and cellular endosperm as a primitive condition, largely because these characters are found in archaic flowering plants, including *Degeneria*.

It is also important to mention, that in the most archaic magoleanean familes, including Degeneriaceae, Magnoliaceae, Eupomatiaceae, Annonaceae, Winteraceae, Illiciaceae, Schisandraceae and Chloranthaceae, the endosperm is cellular. Together with the above-mentioned statistical data, this confirms the primitiveness of the cellular endosperm.

8.3. Types of Endosperm Specialization

There are two main types of the specialization of endosperm —rumination and the development of haustoria. The outer surface of the ruminate endosperm tissue is irregularly ridged and furrowed to varying degrees, often very deeply. This furrowing occurs in a number of magnoliopsid families, especially in Magnolianae (including *Degeneria, Eupomatia,* Annonaceae, *Cinnamosma,* Myristicaceae, *Austrobaileya,* some Aristolochiaceae), and in some liliopsids (some genera of Dioscoreales, Cyclanthaceae, and Arecaceae) (Tamamschian 1951; Periasamy 1962; Corner 1976). Rumination is due to irregular growth activity of the seed coat, or the endosperm itself, during later stages of seed development (Boesewinkel and Bouman 1984). According to Vijayaraghavan and Prabhakar (1984:343), ruminate endosperm could be an ancestral character still occurring in present-day seeds, belonging to both primitive and advanced taxa.

Another and more remarkable type of endosperm specialization is the formation of endosperm haustoria. The haustoria may arise at the chalazal or micropylar end, or at both ends of the developing endosperm. Endosperm haustoria are especially characteristic for taxa that develop the cellular type of endosperm. In the most archaic groups of flowering plants, including Magnoliaceae, endosperm haustoria are usually absent. In those rare cases, when haustoria are present in them, as in *Magnolia obovata* and in Saururaceae, they are chalazal.

Endosperm haustoria evolved independently in various lines of angiosperm evolution. The presence or absence of haustoria is a taxonomically useful embryological character, but the evolutionary trends in endosperm haustoria are not yet well known.

8.4. The Seed Coat

The seed coat is formed by the integuments, the chalazal and raphal tissue, and often also by the included nucellar tissues (Boesewinkel and Bouman 1984). The seed coat, developed from a bitegmic ovule, consists of two layers, which Corner (1976) named testa (from outer integument) and tegmen (from inner integument). The seed coat developed from a unitegmic ovule is called testa. The scar on the seed coat resulting from the disconnection of the seed from the funicle or placenta (in sessile seeds) is called hilum. In anatropous ovules, in which a prolongation of the funicle is fused to the integument, the fused part of the funicle forms the raphe, a characteristic ridge running along the seed and ending at the chalaza.

Usually the seed coat has a very complicated histological structure and consists of distinct layers of various types of cells. Not only the families and genera but, very often even the species within the same genus are distinguished by the structure of seed coat. The study of the seed coat anatomy has, in many cases, great significance in evolutionary classification of flowering plants (Netolitzky 1926; Corner 1976). The seed coat structure has also a definite significance in establishing the evolutionary grade of various groups of the angiosperms. The high structural diversity of surface characters of the seed coat also provides most valuable criteria for the classification (see especially Barthlott 1981).

Seeds of dehiscent fruits are generally more primitive than those of indehiscent ones. Their seed coat is generally less reduced and specialized. The most primitive type of seed coat anatomy is found in Magnoliales and related orders. They are frequently multiplicative (Corner 1976), that is, the cells of integuments divide after fertilization and form more cell layers by periclinal divisions. Therefore, typical magnolealean seeds, like those of *Degeneria* and *Magnolia,* are relatively massive and with complicated seed coats. The *Magnolia* type of seed coat is evidently the nearest to the seed coat of the original magnoliophytes (Zazhurilo 1940; Takhtajan 1948). Although in *Magnolia* and related genera, the tegmen is not multiplicative and usually eventually shriveled and crushed, their testa is strongly multiplicative to form an attractively colored and nutritious sarcotesta and a protective woody endotesta, composed of lignified cells (Netolitzky 1926; Earle 1938; Zazhurilo 1940; Takhtajan 1948; Kapil and Bhandari 1964; Corner 1976; Melikian 1988). This differentiation of the seed coat into sclerotesta and sacrotesta is a typical adaptation for endozoochory. The earliest flowering plants and their immediate ancestors were endozoochorous, probably endozaurochorous (Zazhurilo 1940).

The primitive character of the seed coat of *Magnolia* type is confirmed also by a comparison with the seed coat of a number of archaic gymnosperms, where the outer parenchyma is also connected with an adaptation to endozoochory. In Cycadales, for example, the outer fleshy layer (sarcotesta), which also contains plenty of nutrients and is colored, is formed from the outer envelope (cupula) while the middle woody layer (sclerotesta) is formed from the outer layer of the integument itself. The nearest ancestor of the flowering plants probably had a similar seed coat structure.

In many lines of angiosperm evolution, particularly in plants with indehiscent fruits, a gradual simplification of seed coat is observed. The maximum simplification of the seed coat is attained in those cases where the seed adjoins closely or is fused with the pericarp, as, for example, in Poaceae. The role of protection of the embryo here passes over to the pericarp and, as a result, the seed coat is very reduced and obliterated. In some cases, the reduction of the seed coat goes very far. Frequently, only the outer epidermis of the testa is retained in the mature seed, as, for example, in Apiaceae. In families with ategmic ovules, as, for example, Balanophoraceae, Misodendraceae, Loranthaceae, or Viscaceae, the seed coat does not develop at all.

8.5. *Origin of the Fleshy Seed Appendages*

During the evolution of zoochory, starting from the primitive endozaurochory and ending in the most highly specialized forms of myrmecochory, various types of succulent nutritive appendages of the seed coat and funicle play a major role. At first, presumably, the sarcotesta served as the bait for attracting the animal dispersing agents. It was suggested by Zazhurilo (1940) and van der Pijl (1955) that the first dispersing agents of the earliest flowering plants were the arboreal reptiles, which later were replaced by birds. Sarcotestal seeds were well adapted for dispersal by both the reptiles and birds.

In many flowering plants, the sarcotesta is replaced by special fleshy and edible local outgrowths of the seed or funicle (see Baillon 1876; Netolitzky 1926; Ulbrich 1928; Komar 1965; van der Pijl 1969; Endress 1973; Corner 1976; Kapil et al. 1980). This replacement of the sarcotesta by local fleshy appendages van der Pijl explains in the following way: "In the rain forest, the seeds of *Magnolia* and the like soon lose their power of germination, a weakness which seems unimportant there for climax forms. For other plants, it obviously became important to separate the two functions of attraction and hardness; these should no longer be manifest in two layers as before, but in two parts side by side. Then the juicy part can, moreover, be more easily separated after the transport" (1969:112–113).

All kinds of pulpy structures which grow from some part of the ovule or funicle are known under a general term aril. Traditionally, arils are subdivided on the basis of their point of attachment and origin. Arils originated independently in various lines of flowering plant evolution. They are found already in magnoleanean families Annonaceae and Myristicaceae and are widely distributed in a number of other groups, especially in tropical and subtropical plants. There are many varieties of arils in the most advanced flowering plants, including caruncle, a small disklike structure restricted to the exostome rim around the micropyle (*Ricinus, Euphorbia, Polygala*) and strophiole, a glandular or spongy appendage limited to the raphal region (*Chelidonium, Corydalis, Euonymus, Acacia* spp.).

A special kind of arils, termed elaiosomes, are found in the mirmecochorous plants. The elaiosomes contain an unsaturated, free fatty acid, which attracts ants (Bresinsky 1963). Elaiosomes are small and even diminutive and are found on various parts of seeds and funicles. Arils of *Asarum, Helleborus, Stylophorum, Chelidonium, Dendromecon, Corydalis, Moehringia, Stellaria, Primula, Viola, Reseda, Ricinus, Mercurialis, Euphorbia, Polygala, Scilla, Gagea, Galanthus, Luzula,* and many other plants are elaiosomes. The basis of the elaiosome is often caruncle, but in *Primula* and some species of *Melampyrum* and *Veronica*, it arises from a swelling of the funicle, and—in *Pedicularis sylvatica* from a protruding endosperm haustorium (van der Pijl 1969).

The aril is rare in the most advanced groups of flowering plants.

References

Baillon H. 1876. Sur l'origine du macis de la muscade et des arilles en général. Adansonia 11:329–340.

Barthlott W. 1981. Epidermal and seed surface characters of plants: systematic applicability and some evolutionary aspects. Nord. J. Bot. 1:345–355.

Baude E. 1956. Die Embryoentwicklung von *Stratiotes aloides* L. Planta 46:649–671.

Boesewinkel F.D and F. Bouman. 1984. The seed: structure. In B. M. Johri, ed., Embryology of angiosperms, pp. 567–610. Berlin.

Bresinsky A. 1963. Bau, Enwicklungsgeschichte und Inhaltsstoffe der Elaiosomen. Bibl. Bot. no. 126.

Brink R. A and D. C. Cooper. 1947. The endosperm and seed development. Bot. Rev. 13(8):423–477, 13(9):479–541.

Corner E. J. H. 1976. The seeds of dicotyledons. Cambridge.

Coulter D. M. and C. J. Chamberlain. 1903. Morphology of the angiosperms. New York.

Crété P. 1956. Lentibulariacées: Dévelopment d l'embryon chez *Pinguicula leptoceras*. Rchb. Compt. Rend. Acad. Sci. Paris 242:1063–1065.

Eames A. J. 1961. Morphology of the angiosperms. New York.

Earle T. T. 1938. Origin of the seed coats in *Magnolia*. Amer. J. Bot. 25:221–222.

Endress P. K. 1973. Arils and aril-like structures in woody Ranales. New Phytol. 72:1159–1171.

Grushvitski I. V. 1961. The role of underdevelopment of embryo in the evolution of flowering plants. Moscow and Leningrad (In Russian.)

Haccius B. 1952. Die Embryoentwicklung bei *Ottelia alismoides* und das Problem des terminalen Monokotylen-Keimblattes. Planta 40:443–460.

Haccius B. 1954. Embryologische und histogenetische Studien an "monokotylen" Dikotylen. I. *Claytonia virginica* L. Ost. Bot. Z. 101:285–303.

Hallier H. 1902. Beiträge zur Morphogenie der Sporophylle und des Trophophylle in Beziehung zur Phylogenie der Kormophyten. Jahrb. Hamburg. Wiss. Anst. XIX, 3. Beiheft, 1–110.

Hallier H. 1912. L'origine et le système phylètique des Angiosperms exposés à l'aide de leur arbre généalogique. Arch. Néerl. Sci. Exact. Nat. 3d ser., 1:146–234.

Hegelmaier F. 1874. Zur Entwicklungsgeschichte monokotyledoner Keime nebst Bemerkungen über die Bildung der Samendeckel Bot. Z. 39:631–639, 40:648–656, 41:657–671, 42:673–686, 43:689–700, 44:705–719.

Hegelmaier F. 1878. Vergleichende Untersuchungen über Entwicklung dikotyledoner Keime mit Berücksichtung der pseudo-monokotyledonen. Stuttgart.

Henslow G. 1893. A theoretical origin of endogens from exogens through self-adaptation to an aquatic habit. J. Linn. Soc. 29:485–528.

Henslow G. 1911. The origin of monocotyledons from dicotyledons, through self-adaptation to a moist or aquatic habit. Ann. Bot. 26:717–744.

Kapil R. N. and N. N. Bhandari. 1964. Morphology and embryology of *Magnolia*. Proc. Nat. Inst. Sci. India. 30:245–262.

Kapil R.N., J. Bor, and F. Bouman. 1980. Seed appendages in Angiosperms. Bot. Jahrb. Syst 101(4):555–573.

Komar G. A. 1965. The arils, their nature, structure, and function. Bot. Zhurn. (Leningrad) 50(5):715–724 (In Russian.)

Levina R. E. 1987. Morphology and ecology of fruits. Leningrad. (In Russian.)

Martin A. C. 1946. The comparative internal morphology of seeds. Amer. Midl. Nat. 36:513–660.

Melikian A. P. 1988. Magnoliaceae. In A. Takhtajan, ed., Comparative anatomy of seeds, 2:11–17. (Leningrad). (In Russian.)

Metcalfe C. R. 1938. An interpretation of the morphology of the single cotyledon of *Ranunculus ficaria* based on embryology and seedling anatomy. Ann. Bot. 50:103–120.

Nägeli C. W. 1884. Mechanische - physiologische Theorie der Abstanmungslehre. Munich and Leipzig.

Netolitzky F. 1926. Anatomie der Angiospermen-Samen. In K. Linsbauer, ed., Handbuch der Pflanzenanatomie, Band 10, Lief. 14. Berlin.

Periasamy K. 1962. The ruminate endosperm. Development and types of rumination. In Plant embryology: a symposium, pp. 62–74. New Delhi.

Pijl L. van der. 1955. Sacrotesta, aril, pulpa, and the evolution of the angiosperm fruits. I, II. Proc. Ned. Acad. Wet. (C) 58:307–312.

Pÿl L. van der. 1969. Principles of dispersal in higher plants. New York.

Pritzel E. 1898. Der systematische Welt der Samenanatomie, insbesondere des Endosperms bei Parietales. Bot. Jahrb. 24:345–394.

Rao V. S. 1938. The correlation between embryo type and endosperm type. Ann. Bot. 2:535–536.

Schnarf K. 1929. Embryologie der Angiospermen. In K. Lindsbauer, ed., Handbuch der Pflanzenanatomie, II Abt., 2 Teil. Band 10. Berlin.

Schnarf K. 1931. Vergleichende Embryologie der Angiospermen. Berlin.

Schürhoff P. N. 1926. Die Zytologie der Blutenpflanzen. Stuttgart.

Sporne K. R. 1954. A note on nuclear endosperm as a primitive character among dicotyledons. Phytomorphology 4(3,4):275–278.

Sporne K. R. 1967. Nuclear endosperm: an enigma. Phytomorphology 17:248–251.

Sporne K. R. 1980. A re-investigation of character correlations among dicotyledons. New Phytol. 85:419–449.

Swamy B. G. L. and P.M Ganapathy. 1957. On endosperm in dicotyledons. Bot. Gaz. 119:47–50.

Swamy B. G. L. and N. Parameswaran. 1963. The helobial endosperm. Biol. Rev. 38(1):1–50.

Takhtajan A. L. 1948. Morphological evolution of angiosperms. Moscow. (In Russian.)

Tamamschian S. G. 1951. Rumination of endosperm in angiosperms. Bot. Zhurn. 36(5):497–514. (In Russian.)

Ulbrich E. 1928. Biologie der Frucht und Samen. Berlin.

Vijayaraghavan M. R. and K. Prabhakar. 1984. The endosperm. In B. M. Johri, ed., Embryology of angiosperms, pp. 319–376. Berlin.

Winkler H. 1931. Die Monokotylen sind monokotylen. Beitr. Biol. Pfl. 90(1):29–34.

Wunderlich R. 1959. Zur Frage der Phylogenie der Endospermtypen bei den Angiospermen. Ost. Bot. Z. 106:203–293.

Yakovlev M. S. 1946. The monocotyledonous character in the light of embryological data. Sov. Bot. 14(6):351–362. (In Russian.)

Zazhurilo K. K. 1940. On the anatomy of the seed coats of the Magnoliaceae (*Liriodendron tulipifera* L.). Byull. Obshch. estestvoispyt. pri Voronezhskom Univ. 4(1):32–40. (In Russian.)

9

Mosaics of the Evolutionary Trends and Heterobathmy of Characters

In the course of evolution, correlative interconnections of different parts, organs, and systems of organs of an organism are observed. These evolutionary correlations are expressed in different degrees. When the function of a given organ is necessary for the function of another organ, there is a clear evolutionary correlation between them. For instance, such correlations are evident between the conducting system of leaf petiole and that of the stem or between xylem and phloem. But in cases when such functional interconnections are absent, as, for example, between pollen grains and fruits or between sporophyte and gametophyte, we do not observe any evolutionary correlations. The pollen grains and fruits or the sporophyte and the gametophyte are members of different correlation chains.

Both the paleontological record and comparative morphology of living organisms give strong evidence confirming the relative independence and unequal rates of evolution of many organs and their parts. They clearly show that, while some organs and structures evolve faster, others evolve more slowly, and still others may remain for a long time at a relatively primitive level and in sharp contrast to all other parts of an organism. They also evolve faster in one lineage than another and more rapidly at some times than others. The difference in rates of evolution of different structures in the same phylogenetic lineage was noticed long ago and is known under various names (which, however, are not always complete synonyms). The Belgian paleontologist Dollo (1893) called this phenomenon "chevauchement des specialisations," Eimer (1897) called it "heteroepistasy," and De Beer (1954) "mosaic evolution." In botany, it was formulated by Arber and Parkin (1907) under the title "the law of corresponding evolutionary stages." The Russian botanist Kozo-Poljanski (1940) called it "the law of heterochrony of characters." In my opinion, it is useful to distinguish between the process of mosaic evolution and its result. Therefore, I proposed for the result of mosaic evolution the term "heterobathmy" (from greek *bathmos*-step) (Takhtajan 1959).

In heterobathmic organisms along with more or less primitive (plesiomorphic) characters we observe also more advanced (apomorphic) characters. A heterobathmic organism represents a "disharmonic" combination of characters of different (sometimes very different) evolutionary grades.

The study of heterobathmy in various systematic groups of higher plants leads me to the conclusion that the degree of heterobathmy depends on the degree of evolutionary correlations of the structures, the mode of adaptive evolution, and the evolutionary grade of a given taxon (Takhtajan 1946, 1959, 1966). The differences in the rates of evolution of different structures of an organism first of all depends on the degree of correlations. In modular organisms, such as higher plants, in contrast to the unified organisms, such as a majority of animals, correlative bonds between different parts are of a much lesser degree both in ontogeny and evolution. It is therefore not surprising that heterobathmy of characters is much more frequent in plants than in animals.

The more closely the characters are interlinked during the evolution, the less pronounced is their heterobathmy. Whenever there is a complete evolutionary correlation, the phenomenon of heterobathmy is no longer observed. Therefore, we observe the best examples of heterobathmy in characters that are not interlinked through correlations. In this respect, differences in the evolutionary rates of sporophytes and gametophytes, and in many instances also of vegetative and reproductive organs of the sporophyte itself, are remarkable. As for the conducting systems of different parts of the sporophyte (both vegetative and reproductive), the correlation bonds between their elements are much more strong and the heterobathmy is therefore considerably less pronounced.

However, the degree of heterobathmy depends not only on the degree of correlation bonds. An analysis of heterobathmy in various systematic groups of both extinct and extant higher plants leads to the conclusion that in each large group heterobathmy is relatively more strongly marked in the most archaic members (Takhtajan 1946, 1959, 1966; see also Davis and Heywood 1963:34). In flowering plants, for instance, heterobathmy is most clearly expressed in such archaic taxa as Magnoliales, *Eupomatia,* Winteraceae (especially *Takhtajania*), Chloranthaceae (especially *Sarcandra*), Nymphaeales, Trochodendrales, and others. Among the best examples of heterobathmy are *Takhtajania, Trochodendron,* and *Tetracentron,* which possess very primitive wood devoid of vessels but yet have relatively advanced flowers. *Trochodendron* and *Tetracentron* have also advanced (tricolpate) pollen grains. Well-known examples of heterobathmy are also *Delphinium* and *Aconitum,* which have zygomorphic flowers (apomorphic character) together with an apocarpous gynoecium (plesiomorphic character). On the contrary, in the most advanced groups of flowering plants, such as

Lamiaceae, Asteraceae, Orchidaceae, or Poaceae, heterobathmy is usually expressed to a relatively very slight degree. We observe also an analogous phenomenon in other groups of higher plants.

The basic reason for more frequent and stronger heterobathmy in the archaic groups lies in their evolutionary primitiveness as such. Heterobathmy is especially obvious at the earlier evolutionary stages of new progressive lineages. It is probably explained by the fact that the origin of any new large phylogenetic branch is usually associated with the disintegration of previous correlation chains and the establishment of new ones. Such a profound reconstruction of the correlations takes place both in progressive and regressive evolutionary lines. In both cases, not all parts of the organisms are reconstructed at the same time, at the same rate, and to the same degree, which inevitably produces the heterobathmy of characters.

In the process of subsequent evolution of heterobathmic groups, the degree of heterobathmy usually tends to decrease. This progressive equalization of the evolutionary grades of different characters of one and the same group of organisms usually results from the increasing evolutionary specialization. Therefore, while the archaic forms are characterized by "disharmonic" combinations of heterobathmic characters, the more advanced forms have more equalized grades of their different parts. One can explain this progressive alignment of advancement levels by the increasing approach of more and more structures to their highest degree of specialization. Evidently, in each given evolutionary trend, there is a certain relatively highest level of development or a culmination stage of specialization at which the structure is best adapted to the execution of its function in a given set of environmental conditions.

The decrease of heterobathmy, the "harmonization" of the combination of different characters belonging to different evolutionary series, is also very much favored by the establishment of broader adaptive correlations to some special environmental conditions. A good example is an adaptation to arid climate of deserts. Xerophilization is one of those modes of specialization, which favors to a high extent the leveling of heterobathmy. Therefore, in xerophytes we rarely observe those clearly expressed cases of heterobathmy, which are found in many mesophytes.

In many cases of evolutionary degradation, especially in cases of simplification and reduction of only some parts of an organism, heterobathmy is increased. The most vivid examples of heterobathmy connected with partial degeneration are found in many aquatic plants, saprophytes, and parasites. One of the best examples is the genus *Cassytha,* in which the reproductive organs are still on the level of organization of other autotrophic lauraceous genera, while the vegetative organs are both extremely simplified and specialized. Other analogous examples are various genera of the order Rafflesiales, such as *Rafflesia, Cytinus,* and others. No less pronounced is the heterobathmy between the reproductive and vegetative organs in aquatic plants. However, in those cases in which degeneration affected all parts of the sporophyte, the heterobathmy again decreases, as for example in the Lemnaceae.

Thus, we come to the conclusion, that heterobathmy depends on the degree of correlation between different parts of an organism, as well as on the evolutionary grade and mode of adaptive evolution of a given systematic group.

References

Arber E. A. N. and J. Parkin. 1907. On the origin of angiosperms. J. Linn. Soc. London, B 38:29–80.
De Beer G. R. 1954. Archaeopterix lithographica. London.
Davis P. H. and V. H. Heywood. 1963. Principles of angiosperm taxonomy. Edinburgh and London.

232 *Mosaics of Evolutionary Trends*

Dollo L. 1880. 1893. Les lois de l'évolution. Bull. Soc. Belg. Geol. 7:164–166.

Eimer Th. 1897. Die Entstehung der Arten. I-II. Leipzig.

Kozo-Poljanski B. M. 1940. The laws of plant phylogeny and Darwinism. In B. A. Keller, ed., Plants and environment, pp. 43–66. Moscow and Leningrad. (In Russian.)

Takhtajan A. 1946. On the evolutionary heterochrony of characters. Doklady Armenian Acad. Sci. 5:79–86 (In Russian with English summary.)

Takhtajan A. 1959. Die Evolution der Angiospermen. Jena.

Takhtajan A. 1966. Systema et phylogenia magnoliophytorum. Moscow and Leningrad. (In Russian.)

Index

Abbreviation, 4, 5, 7, 185, 186, 190: basal, 5-7, 186, 192; medial, 5-7; terminal 4-9, 186, 192
"Abort-Hypothese," 214
Acceleration of development, 185, 186, 192, 196
Achenium, 202, 203
Acorn, 206
Additions, 3
Aestivation, 81, 82; contorted, 82; convolute, 82; imbricate, 81, 82; induplicate, 82; plicate, 82; quincuncial, 82; valvate, 82
Albedo, 209
Albuminous cells, 58-59
"Alternanz-Regel," 87
Ament, 121
Anaboly, 3, 14
Androecium, 79, 86-90; centrifugal, 89; centripetal, 88; cyclic, 87-89
Androecium,; diplostemonous, 87, 88; haplostemonous, 87, 88; obdiplostemonous, 87; obhaplostemonous, 87; oligomerization of, 88; polymerization of, 88; polymerous, 78, 88; polystemonous, 87; secondary polymery, 85; spiral, 87, 88; splitting of primordia, 88

Andropetals, 80-81
Anemochory, 202
Anemophily, 178, 179; origin of, 178-179
Anthela, 121
Antheridium, 188
Anther(s), 83-86; abaxial position, 85; adaxial position, 85; basifixed, 86; endothecium, 135, 136; fibrous layer, 135; immersion of, 84; longitudinal dehiscence, 137, 204; poricidal dehiscence, 137, 205, 206
Anther(s),; protection in cantharophilous flowers, 84; transverse dehiscence, 138; valvate dehiscence, 137; versatile, 86
Anthotaxis, 77
Antipodals, 189
Aperture, 146-148, 149, 153; compound, 152; distal, 147-151; global, 147, 150; nonorate, 153; orate, 152, 153; simple, 152; zonal, 147, 149
Apocarpous fruits, 200, 206
Apocarpous gynoecium, 96, 97-98, 100
Aporogamy, 194
Arborescent forms, secondary, 25

Arborescent liliopsids, 25
"Archaism of bases," 15
Archallaxis, 6, 7, 85
Archegonial disappearance theory, 191-192
Archegonium, 191, 192
Archiboly, 6
Aril, 222-223, 231; micropylar, 161, 189, 216
Autochory, 201
Axial parenchyma, 63, 64; apotracheal, 63-65; boundary, 65; diffuse, 63-64; marginal, 65; metatracheal, 65; paratracheal, 65; vasicentric, 65
Axile placentation, 103, 204

Bacca, 209
Balausta, 206
Berry, 208, 209; inferior, 209; paracarpous, 206-210; syncarpous, 208-210
Bisamara, 205
Bitegmic, 161-162, 163, 164, 167, 220
Body cell, 186
Botryose infloresences, 113
Bracteoles, 119, 121
Bracteopetals, 79, 80, 81
Branching, 26-27; monopodial, 26, 27; original type, 26; sympodial, 26, 27

Calyx, 79, 81, 82, 87
Cantharophily, 84
Capitulum (capituli), 122; cymose, 122, 123; racemose, 123
Capsule, 204-209; circumscissile dehiscence, 205; dorsicidal dehiscence, 204; inferior, 205-207, 209; loculicidal dehiscence, 205; paracarpous, 204, 206-210; poricidal dehiscence, 205, 206; septate, 204-207; septicidal dehiscence, 205; syncarpous, 206, 209-210; unilocular with free-central placentation, 204; uni-

locular with parietal placentation, 204
Carpel(s), 89-108; concrescence of the margins, 93; conduplicate, 92-93, 95-96; dorsal vein, 89, 96; free margins, 89; fusion of lateral veins, 93; incomplete fusion of margins, 93; origin, 90, 91, 92; phyllome character, 90; primitive type, 92-93
Carpel(s),; sealed, 93, 100; stipitate, 92; venation, 92; ventral veins, 92, 96
Carpidium, 200
Carpophore, 210
Caruncle, 223
Caryopsis, 207, 210
Catkin, 121
Chalaza, 194, 219
Chalazal cell, 189, 216
Chalazogamy, 194
Characters, 10-16; advanced, 11-14, 16; apomorphic, 229; correlations of, 22, 234; evolutionary sequence, 12, 13, 14, 15, 16; evolutionary series, 12; inversions of the evolutionary sequence, 15-16; heterobathmy, 228-231; morphoclines, 12; plesiomorphic, 229, 235; polarity, 12, 14; transformation series, 12
Cynarrodium 203
Cladogenesis, 11
Coenobium, 205
Coenomegaspore, 190
Colpus, 147-149
Columellae, 144-146
Columnar placentation, 103, 105
Companion cells, 59
Compitum, 98
Connective, 83; protrusion of, 83, 86
Corolla, 79, 81, 82, 90; choripetalous, 81; evolutionary specialization, 81, 90; origin, 79-81; sympetalous, 81
Corymb, 121, 122

Cotyledon(s), 214-215
Crassinucellate, 166, 167
Cremocarp, 210
Cross pollination, 113, 171
Cryptotetrads, 139
Cupule, 162, 165, 221
Cyclocolpate pollen grains, 149
Cymose inflorescences, 113-115
Cypsela, 207

Decurrent stigma, 94, 96
Deviation, 6, 7; basal, 4-7; medial,
    4-7; terminal, 3-7; total, 6, 7
Dichasial inflorescences, 121-122
Dichasium, 115-118; capitate, 117;
    compound, 116-117; corym-
    bose, 117; simple, 115-117; um-
    belliform, 117, 118
Dicolpate pollen grains, 151
Diffuse placentation, 105-107
Dorsal placentation, 105
Double fertilization, 194-195
Drupe, 199, 203-204; inferior, 209-
    210; lysicarpous, 207-208; mon-
    opyrenous, 208; polypyrenous,
    208; syncarpous, 208, 209, 210
Dyads, 139

Ectexine, 141-143
Elaiosome, 223
Embryo, 213-215, 217, 218, 221;
    dicotyledonous, 214, 215; in-
    creased size of, 214; monocotyle-
    donous, 214, 215, 232; nonde-
    velopment of one of the two co-
    tyledons, 215; origin of monoco-
    tyledonous embryo, 215; undif-
    ferentiated, 213, 217, 218
Embryo-sac, 188
Endexine, 141, 143
Endocarp, 203, 209, 210
Endozaurochory, 221, 222
Endosperm, 195, 213, 216-219; cel-
    lular, 192, 216-219; helobial,
    216; nuclear, 216-218; reduc-
    tion, 216; ruminate, 219
Endosperm haustoria, 219

Endotesta, 221
Endothecium, 135
Endothelium, 167
Endozoochory, 201, 203, 208, 221
Entomophily, 172, 174, 178, 179
Eremus, 205
Exine, 140-148; atectate, 143; intec-
    tate, 146
Exine,; perforate, 146; semitectate,
    146; tectate, 144; tectate-imper-
    forate, 144
Exocarp, 203, 209, 210
Exostome, 223

False septum, 205, 206, 207
Fiber-tracheids, 65, 67
Filament, 83, 86, 89-90
Flavedo, 209
Floral axis, 76, 199, 200
Floral tube, 176, 209
Flower, 75-82; cyclic, 76, 77; dis-
    tinction from the gymnosperm
    strobil, 75; fixation of the num-
    ber of parts, 76; oligomerization,
    78, 88; origin, 75, 76; polymeri-
    zation, 78; primitive types, 78,
    80-81; solitary, 115, 118, 116;
    spiral, 76, 77, 82; spirocyclic,
    77; structural integration, 78;
    terminal arrangement, 113-
    114
Flower color, evolution, 173, 174
Follicle, 200, 201, 203
Foot-layer, 144
Fragum, 203
Free central placentation, 100, 103,
    105, 109
Fruits, 199-210; apocarpous, 200,
    206; classification, 199, 200;
    dry, 205, 206, 210; dry syncar-
    pous, 204; fleshy, 199, 201,
    203; fleshy syncarpous, 208-210;
    lysicarpous, 207; paracarpous,
    206-210; syncarpous, 200, 202
Fruitlets, 200-204
Fruit wall, 199
Funicle, 164, 165, 220, 222, 223

Gametangia, 185
Gametes, 185, 186, 195
Gametogenesis, 185, 186, 191
Gametophytes, 185, 186, 188, 189, 190, 191, 192, 194, 195; female, 190-195; female, Adoxa-type, 190; female, antipodal cells, 190; female, bisporic, 190; female, chalazal end, 189, 194
Gametophytes,; female, elimination of archegonia, 191; female, egg apparatus, 189, 194; female, gametangialess, 185; female, monosporic, 189, 190; female, neotenous origin, 195; female, *Oenothera*-type, 190; female, *Penaea*-type, 195; female, *Peperomia*-type, 195; female, *Plumbago*-type, 195; female, polar nucleus, 194, 195; female, *Polygonum*-type, 189, 190; female, secondary nucleus, 190, 194; female, synergids, 189; female, tetrasporic, 190, 192, 195; male, 185-188; male, generative cell, 186, 187; male; origin of, 185-188; male, tube cell, 186
Generative cell, 186-188
Glans, 206
Growth habit, 21
Gynoecium, 96-109; apocarpous, 96-105; coenocarpous, 97; cyclic, 95-97, 99; eusyncarpous, 99-102
Gynoecium,; lysicarpous, 99, 103; monomerous, 97; paracarpous, 99-103; pseudomonomerous, 103-104; pseudosyncarpous, 104; reduction in the number of carpels, 101, 103; spiral, 98, 99; syncarpous, 97-101, 103

Head, 121, 122
Helicoid cyme, 118
Herbaceous stem, 26
Hesperidium, 209

Heterobathmy, 227-231
Hilum, 220
Hypocotyl, 214
Hypsophylls, 79

Inaperturate pollen grains, 148, 152
Inframarginal vein, 34, 40, 41
Infratectal layer, 143, 144; columellar, 143-146; granular, 143-145
Inflorescences, 112-123; aments, 121, 123; anthela, 121
Inflorescences,; botryose, 113; capitate dichasium, 117; capitate raceme, 121; capitulum, 122, 123; catkin, 121; centrifugal, 113; centripetal, 113; closed, 113, 115; composite, 123, 124; compound dichasium, 116, 117; compound monochasium, 117, 118; compound umbel, 123; corymb, 121, 122; corymbose dichasium, 117; cymose, 113-115, 119, 121, 123; semose head, 121; cymose panicle, 115, 118, 119; determinate, 113, 114; dichasial inflorescences, 115-117, 119; dichasium, 114-117; head, 117, 121, 122; helicoid cyme, 118; intermediate, 113-114, 119; monochasial, 117, 118; monochasium, 117, 118
Inflorescences,; monotelic, 113, 114, 119; open, 113, 114; origin, 112-113, 123; panicle, 115, 118, 119; polytelic, 113, 114, 118; primitive, 112, 113, 115; raceme, 114, 118-122; racemose, 113-115, 118-123; racemose umbel, 121-122; rhipidium, 118; scorpioid cyme, 117; secondary character of solitary arrangement, 112; simple dichasium, 115, 117; spadix, 121; spike, 121; syconium, 117; thyrse, 123; umbel, 118; umbel-

liform monochasium, 118; umbelliform dichasium, 117; umbellum, 122
Integument, 160-164; cupular origin of the outer integument,
162; reduction of integument,
164; segmentes in archaic gymnosperms, 160-161; synangial
hypothesis of the origin of integument, 160
Integumentary tapetum, 167
Intine, 140-142, 146-148

Juvenilization, 8

Laminar diffuse, 105-107
Laminar dorsal, 105
Laminar lateral, 105-107
Laminar placentation, 94, 104-106
Law of integration of the homologues, 78
Leaves, 28-45; alternate, 45; compound, 29, 30, 32; entire, 29,
30, 39; opposite, 45, 56; palmately compound, 30; palmately
lobed, 30; pinnate, 29, 30, 32,
34, 35, 38-41; pinnately compound, 30; pinnately lobed, 30;
primitive type, 39-41, 45; simple, 28-30, 32, 33, 39, 40; verticillate, 45
Legume, 201, 202; indehiscent,
201, 202; jointed, 202; nutlike,
202; succulent, 201, 202
Libriform fibers, 65, 67
Living fossils, 10-11
Loment, 202
Lumina, 146
Lysicarpous gynoecium, 103, 105

Marginal placentation, 93
Median placentation, 106, 107
Megasporangium, 159, 160, 164,
167, 194
Megaspore mother cell, 166, 167
Megaspores, 159, 166, 189, 190

Megasporocyte, 167
Megasporophylls, 90, 91, 159
Megasynangium, 160
Mericarps, 205, 210
Mesoboly, 6
Mesocarp, 203, 209, 210
Mesogamy, 194
Mesomes, 161
Metaxylem, 51, 54
Micropylar cell, 216
Micropyle, 159-162, 165, 194
Microsporangium, 136-138; abaxial
arrangement, 85; adaxial ar-
rangement, 85; embedded, 84;
immersion of, 84; laminar ar-
rangement, 84
Microspore mother cell, 138, 139
Microspore tetrads, 138
Microspores, 136, 138, 141, 142,
147, 187
Microsporocyte, 136, 138
Microsporogenesis, 136, 138, 139;
simultaneous, 138, 139; succes-
sive, 138, 139
Microsporophyll, 84
Midrib, 31, 38-39, 92, 206
Miniaturization, 186, 195
Minor veins, 42-45; closed type, 43,
44; open type, 43, 44
Monocolpate pollen grains, 148,
149, 151
Monads, 139
Monochasium, 117, 118
Monoporate pollen grains, 148,
150, 152
Monotelic inflorescences, 113-115,
119
Morphoclines, 12, 13
Mosaic evolution, 227-231
Multiachenium, 202-203; sub-
merged, 203; syncarpous, 204
Multifollicle, 200-203, 204; cyclic,
201, 204; spiral, 200-204; suc-
culent, 201, 203; syncarpous,
204
Muri, 146

Mutual selection, 178
Myrmecochory, 223

Nectaries, 173
Neotenous origin of flowering
    plants, 76
Neoteny, 7-9, 192, 195
Nexine, 142
Nodal structure, 47-50; multilacu-
    nar, 47, 49; pentalacunar, 48;
    trilalunar, 49; unilacunar, 47-49
Nucellar cavity, 218
Nucellus, 159, 160, 165-167
Nut, 206
Nux, 202, 206

Ockham's Razor, 11
Oligomerization, 78, 88
Onci, 142
Ontogeny, 2, 3, 5-9, 14-16; closed,
    14; serial, 14-15
Ora, 153
Orbicules, 137
Ornamentation, 143, 144, 145, 146
Out-group analysis, 13
Ovary, 103, 109-112; adnate, 110;
    free, 109; inferior, 109-112;
    multilocular, 103, 109; superior,
    109; unilocular, 103, 109
Ovule, 159-167; amphitropous,
    164, 166; anacampylotropous,
    166; anatropous, 164-166;
    ategmic, 162, 164; atropous,
    164; bitegmic, 161, 162, 163,
    167; campylotropous, 164, 166;
    circinotropous, 165; crassinucel-
    late, 166, 167; decrease in the
    number of, 107; hemitropous,
    164; origin of unitegmic ovule,
    163; orthocampylotropous, 166;
    orthotropous, 164, 166; simplifi-
    cation, 159; tenuinucellate, 166,
    167; unitegmic, 162-164, 220

Paedomorphosis, 8
Palaeobotanical record, 10, 13
Panicle, 115-116, 118

Papaverella, 206
Paracarpous gynoecium, 99-103
Parietal placentation, 98, 100, 105,
    109
Pepo, 210
Perforation plate, 55, 56; scalari-
    form, 52, 55; simple, 57-60
Perianth, 77-82; differentiation, 79;
    origin of, 77-79; simple, 79
Pericarp, 199, 202, 203, 205-209,
    221
Pericolpate pollen grains, 152
Periporate pollen grains, 150
Perisperm, 213
Petals, 79-82; bracteal origin of, 79;
    dual origin of, 80; staminal origi-
    nal of, 80
Petiole, 38-41
Phylembryogenesis, 2, 3
Phyllotaxy, 45, 50
Placenta, 102, 159, 165
Placentation, 93-109; axile, 103,
    106, 108, 204; columnar, 103,
    105, 109; diffuse, 105-107; dor-
    sal, 105; free central, 100, 103,
    105, 109; laminar, 94, 106, 107;
    laminar-diffuse, 105-107; lami-
    nar-dorsal, 105; laminar-lateral,
    105-107; marginal, 93, 102,
    103; median, 103, 106; parietal,
    98, 100, 105, 109; submarginal,
    91, 104, 105, 107, 108; sutural,
    104, 107
Plumule, 214
Pollen grains, 139-153; cyclocol-
    pate, 149; dicolpate, 151; ina-
    perturate, 148, 152; monocol-
    pate, 148, 149-151; mono-
    porate, 148, 150, 152; pancol-
    pate, 150; panporate, 150;
    pericolpate, 152; polyporate,
    152
Pollen grains,; porate, 150; spiraper-
    turate, 149; three-celled, 140,
    193, 194; trichotomocolpate,
    149; tricolpate, 151; tripoly-
    pororate, 153; triporate, 152;

two-celled, 193; zonocolpate, 151-152; zonoporate, 150, 153; zonopororate, 153
Pollenkitt, 136
Pollen tube, 140, 143, 146, 159, 194
Pollen wall, 140, 146, 156-159
Pollination, 171-173, 175-178; anemophily, 178, 179; cantharophily, 175; coevolution of flowers and pollinators, 176-177; constancy of flower visitation, 180; cross pollination, 113, 171; entomophily, 172, 174, 178, 179; evolution of flower color, 173, 174; hydrophily, 148; insect-trapping flowers, 175; nectaries, 173; origin of anemophily, 178-179
Pollination,; primitive pollinators, 176; secondary entomophily, 172; self-pollination, 81, 171; standardization of the dimensions of flower parts and insect body, 177-178; vectors of pollen transfer, 171; wasp pollination, 176
Pollinium, (pollinia), 139
Polyads, 139
Polyporate pollen grains, 150, 152
Pome, 209
Porate pollen grains, 150
Pore, 136, 150, 152; distal, 148-151; global, 150; zonal, 149
Porogamy, 194
Praefloration, 81
Primary endosperm nucleus, 213, 216
Primitive pollinators, 176
Primordium, 85, 88, 215
Proembryo, 215
Progenesis, 8
Prolongation, 3, 4; basal, 4-7; medial, 4-7; terminal, 3-7
Prothallial cells, 186
Protoxylem, 54
Pseudomonad, 139

Pseudomonomerous gynoecium, 103-104, 205
Pulpa, 206
Putamen, 203, 208
Pyrene, 208
Pyxidium, 205
Pyxis, 205

Raceme, 114, 118-121
Racemose inflorescences, 113-115, 122
Radial parenchyma, 61
Radicle, 214
Raphe, 220
Rays, 61, 62, 63, 67; heterocellular, 61; heterogeneous, 61-62; homocellular, 61; homogeneous, 61; multiseriate, 61, 63; uniseriate, 61, 63
Recapitulation, 13, 14
Receptacle, 75, 76, 88, 89
Regma, 205
Replum, 207
Retardation, 191
Retention, 15; serial, 14-15; transverse, 14, 15
Rhipidium, 118
Ruminate endosperm, 219
Rumination, 218, 219

Samara, 205
Sarcotesta, 221, 222
Schizocarp, 205
Sclerotesta, 161, 221
Scorpioid cyme, 118
Sculpturing, 143, 144, 146, 147
Secondary entomophily, 179
Seed(s), 213-221
Seed appendages, 222-223
Seed coat, 219-221; gradual simplification, 221; primitive type, 220
Self-pollination, 81, 171
Semimericarp, 205
Sieve areas, 58-60
Sieve cells, 58, 59

Sieve elements, 43, 58, 59
Sieve plates, 58-60; compound, 59, 60; simple, 59, 60
Sieve tubes, 58-60
Siliqua, 207
Simplification, 185, 192, 195
Sole, 144
Spadix, 121
Spermatogenous cell, 188
Spike, 121
Spiraperaturate pollen grains, 149
Sporogenous tissue, 136
Sporopollenin, 142-144
Stalk cell, 188
Stamens, 80-91; connective, protrusion of, 83; filament, 83; multisporangiate, 136
Staminal tube, 90
Staminodia, 88, 90
Stigma, 92, 95-96; capitate, 93; decurrent, 94, 95; localized, 92, 95-96; origin of, 94-95; primitive, 95-96
Stigmatic crests, 96
Stigmatic surface, 94-95
Stigmatic tissue, 96
Stomata, 45, 46, 47; actinocytic, 45; anisocytic, 45; anomocytic, 46, 47; diacytic, 45; mesogenous, 46, 47; mesoperigenous, 46; paracytic, 45; perigenous, 46-47; tetracytic, 44
Stone, 208-210
Strophiole, 223
Style, 95, 98, 100
Stylodium, 92, 95, 96; conduplicate, 95-96
Submarginal placentation, 91, 104, 105, 107, 108
Subsidiary cells, 45-47
Sulcus, 147
Sutural placentation, 104, 107
Syconium, 117
Syncarpous fruits, 204, 205, 207, 208; dry syncarpous fruits, 204, 205; fleshy syncarpous fruits, 208, 209

Syncarpous gynoecium, 97-101, 103
Syngamy, 194

Tapetum, 136, 137; amoeboid, 136; glandular, 136; periplasmodial, 136; secretory, 137
Tectum, 143-146
Tegmen, 220
Telome, 161
Telome theory, 161
Tenuinucellate, 166, 167
Tepals, 79, 173
Testa, 220, 221
Tetrads, 138, 139
Tracheary elements, 51, 63-64
Tracheids, 51-58, 65, 67
Transmitting tissue, 96
Trichotomocolpate pollen grains, 149
Tricolpate pollen grains, 151, 152
Triple fusion, 191, 194, 198
Triporate pollen grains, 149
Trophophylls, 91
Trophosporophylls, 91
Tryphine, 136
Tube cell, 186, 187, 194

Umbel, 118, 121
Underdevelopment hypothesis, 215
Unifollicle, 201; succulent, 201, 203
Utricle, 207

Vectors of pollen transfer, 171
Venation, 29-32, 35, 36-42; acrodromous, 35; actinodroma marginalis, 35; actinodromous, 35, 41; arcuate-striate, 36, 41, 42; brochidodroma, 34; camptodrom, 33; campylodromous, 41; coarcuate, 33, 34, 39; collimate, 37; compound rectipinnate, 32; convergent, 35; craspedodromous, 39; curvimarginal, 37; curvipalmate, 35, 40; curvipinnate, 32; dictyodroma, 34; eucamptodromous, 33; longitudinally striate, 37, 41; looped, 33,

39, 40; lyrate, 37, 42; multiarched, 33, 39
Venation,; palinactinodromous, 35;
palmate, 34, 40, 41; palmate-
pinnate, 34, 39, 40, 41; parallel,
34, 37, 41; parallelodroma, 37;
paxillate, 34, 37, 39; pedate, 35,
41; pedate-striate, 36; pinnate,
29-31, 34, 38-41; pinnate-
striate, 36, 42; rectipalmate, 35,
40; rectipinnate, 32, 35, 39; reticulate, 35, 146; reticulipalmate,
35, 40; reticulipinnate, 34; semicraspedodromous, 33, 39; semilooped, 33, 39; simple curvipinnate, 33; simple looped, 33, 39,
40; simple rectipinnate, 32;
striate, 32, 36, 37, 41, 42

Vernation, 45
Vessel elements, 52, 54, 55
Vessels, 52-58; origin of, 52-53;
specialization, 54-56
Vesselless flowering plants, 52

Wasp pollination, 176
Wood fibers, 65
Wood parenchyma, see Axial paren-
chyma, 63, 64, 65, 218
Wood rays, see Rays 61-63

Xylem rays, 61

Zonocolpate pollen grains, 151, 152
Zonoporate pollen grains, 150, 152
Zonopororate pollen grains, 153
Zoochory, 222